Myth Wrapped in Math

Sandeep Chavan

Published by SANDEEP CHAVAN, 2025.

MYTH WRAPPED IN MATH

First edition. August 13, 2025.

ISBN: 979-8231039555

Written by Sandeep Chavan.

Table of Contents

To the mathematicians who dared to dream in numbers,and to the physicists who sought truth beyond them.May this work inspire both to see the stillness beneath the ripples,and the reality beyond the equations.

"Equations can describe the dance,but they cannot hear the music.To find truth, we must step beyond the symbolsand into the stillness that moves them."

Disclaimer

This work is a presentation of the author's independent research, observations, and interpretations under the framework of Universal Energy Dynamics (UED). It is intended for educational, discussion, and exploratory purposes only. While every effort has been made to ensure accuracy and clarity, the concepts described challenge prevailing scientific and philosophical models and may differ from established consensus. Readers are encouraged to critically examine all ideas presented and cross-reference them with credible, peer-reviewed sources.

The content does not claim to represent the official stance of any academic institution, research organization, or governing body. Any historical, scientific, or technical references are provided for context and should not be taken as endorsements. The author assumes no responsibility for how the information is interpreted, applied, or acted upon by readers, whether in academic, professional, or personal contexts.

This material should not be considered a substitute for formal education, professional consultation, or empirical verification. All experimental or observational suggestions are to be undertaken with appropriate expertise, resources, and safety protocols.

By engaging with this work, the reader acknowledges that exploration, debate, and reinterpretation are inherent to the process of seeking truth—and that the ultimate responsibility for discernment rests with the individual.

Opening Note

There is a strange, quiet religion in the modern world — one without temples, without incense, without priests in robes. Its altars are blackboards, its scriptures are equations, and its high priests wear lab coats. The faithful gather not in churches, but in conference halls and peer-reviewed journals. And its most sacred commandment is simple: *If the math works, it must be true.*

But must it?

This book is not an attack on mathematics — it is a rescue mission. Mathematics is one of humanity's greatest tools, a language so precise it can describe the arcs of planets and the structure of atoms. Yet somewhere along the way, we stopped using it as a description of reality and began worshipping it as reality itself.

In ancient times, the myth-maker told stories of gods and chariots of fire to explain the sky. Today, the modern theorist tells stories of singularities, multiverses, and time loops to explain the universe. The language has changed — words replaced by symbols — but the structure is the same: internally consistent, emotionally compelling, and often entirely disconnected from physical truth.

Under **Universal Energy Dynamics (UED)**, the first question is never *Does the equation work?* but *Does the cause-effect match the reality of the ripple field?* If not, then the equation is just another myth — a beautiful one, perhaps, but still a story.

This book will take you through the most celebrated "truths" of modern science — time travel, black hole singularities, dark matter, cosmic inflation — and show how many of them are simply mythology written in mathematics. The goal is not to tear down science, but to free it from the danger of believing the blackboard over the universe itself.

Because belief belongs to myth. And proof belongs to reality.

Prologue

On a humid evening in ancient Babylon, a priest stands on the temple steps, tracing the stars with a bronze-tipped stylus. Around him, villagers gather, waiting for the verdict. His charts are covered in neat lines and symbols. When he speaks, his voice is certain: *"The gods are pleased. The harvest will be good."*

Thousands of years later, in a brightly lit laboratory, a physicist stands before a whiteboard filled with tensors, integrals, and Greek letters. Around him, graduate students watch, waiting for the verdict. His tone is equally confident: *"The equations confirm it — time travel is theoretically possible."*

The gap between these two scenes is not as wide as it seems. Both the priest and the physicist are reading symbols. Both are interpreting them into narratives about a reality no one has directly experienced. Both carry the authority of their craft. And both, without realizing it, are creating worlds that may exist only in the mind.

We like to think science has outgrown mythology, that we've replaced superstition with reason. But in some of the grandest ideas of modern physics, the story has simply changed costumes. Instead of golden chariots, we have warp drives. Instead of divine realms, we have extra dimensions. Instead of prophecies, we have predictions that can't be tested for centuries — if ever.

Mathematics is a language, not a landscape. It can map reality, but it is not reality. A map can tell you where mountains should be, but if you climb and find only empty plains, the map is wrong — not the land. Yet modern science often inverts this logic, defending the equation even when the physical terrain refuses to match.

Universal Energy Dynamics (UED) takes a different stance: reality decides, not the blackboard. If a model doesn't align with actual ripple-field causality — with the measurable chain from alignment to consequence — then no matter how elegant, it's just another myth.

This book is about that dangerous drift — how mathematics, in the absence of physical grounding, becomes modern mythology. We will journey through the celebrated "truths" of contemporary physics, peel away their symbolic elegance, and see which ones survive contact with reality.

Because in the end, a beautifully written myth is still a myth. And the universe doesn't bend for the sake of a good story.

Part I – The Myth-Making Machine

Chapter 1 – When Numbers Became Gods

Long before numbers ruled the cosmos, they counted sheep. They measured sacks of grain, kept track of debts, and marked the days until the next flood. Mathematics was a humble servant — a tool as ordinary as a basket or a plough.

But over centuries, something changed. The simple tally marks of farmers became the geometry of pyramid builders, the star charts of temple astronomers, and the cryptic harmonies of Pythagoras. Numbers stopped being mere *descriptions* of the world and began to be seen as the *essence* of the world.

In time, the farmer's ledger became the priest's prophecy — and eventually, the physicist's equation. The ink and symbols changed, but the authority they carried grew until, in some domains, numbers no longer answered to reality. Reality was expected to answer to them.

This is the story of how the blackboard became an altar, and how "if the math works" replaced "if the gods will it" as the supreme truth.

1.1 From Counting Stones to Cosmic Scripts

The earliest mathematics was born out of necessity, not philosophy.

When the first farmers scratched tally marks into clay tablets, they weren't trying to unravel the mysteries of the universe — they were making sure no one walked away with an extra sack of barley. Counting was survival. Measurement was fairness. Mathematics was a language of trade and trust, rooted in the dust of everyday life.

In the river valleys of Egypt and Mesopotamia, numbers soon found a second purpose: managing the land itself. The Nile's annual flood washed away property boundaries, so geometry emerged to redraw them. This was still mathematics as a servant — practical, grounded, and limited to what could be seen and touched.

Then came the first great leap toward abstraction. Babylonian astronomers began tracking the movements of the stars and planets, not just to admire them, but to predict the seasons and plan religious festivals. The sky became a vast ledger, and mathematics became a way to "read" it. These calculations slowly took on a sacred air: if the patterns in the heavens could be written in numbers, then perhaps the numbers themselves were divine.

In Greece, this idea crystallized into something far more powerful — and far more dangerous. Pythagoras and his followers didn't just use numbers; they worshipped them. They spoke of numbers as the fundamental reality, the invisible scaffolding of existence. Mathematics was no longer merely a tool to measure the world; it was proclaimed as the *source* of the world's order.

This was the birth of a belief that still haunts science today: that numbers don't merely *describe* reality — they *govern* it. Once this seed was planted, mathematics was no longer chained to observation. It could wander, extrapolate, and invent entire worlds without ever touching the soil of physical experience. The transition was subtle but profound: the ledger had become a scripture, and the counter had become a prophet.

1.2 The New Priests of the Blackboard

In ancient civilizations, the most trusted figures were not kings or warriors, but the interpreters of hidden patterns. The Babylonian priest who studied the stars, the Egyptian scribe who read the flood levels of the Nile, the Mayan astronomer who charted Venus — all held the power to declare the will of the cosmos. Their tools were clay tablets and star charts, their authority resting on a mix of skill, tradition, and public awe.

The pattern is easy to miss, but it is still with us today. The temple steps have been replaced by university lecture halls, the clay tablets by blackboards and digital screens. The interpreters are no longer called priests, but physicists, cosmologists, and mathematicians. Their charts have changed form — tensor equations instead of constellations — but the public's relationship to them remains eerily familiar: trust the interpreter, for you cannot read the signs yourself.

In both cases, the process is the same:

1. Gather observations.

2. Translate them into a symbolic system (mythic imagery or mathematical notation).
3. Derive a meaning or prediction.
4. Present it as truth to those who cannot follow the symbolic reasoning themselves.

The key difference is in the language, not the method. Ancient star charts used pictograms of lions and scorpions. Modern physics uses symbols like Σ, π, and c. Both produce a result that only the initiated can "read," and both require the audience to take the conclusion on trust.

Consider a case study:

- **Ancient Eclipse Prophecy** – A Mesopotamian priest observes the moon's position relative to certain stars. Based on recurring cycles recorded on clay tablets, he predicts an eclipse will occur next season. When it happens, the public hails him as the voice of the gods.
- **Modern Gravitational Wave "Detection"** – A physicist examines data from LIGO, a machine sensitive enough to measure distortions smaller than a proton's width. The output is filtered through statistical algorithms and matched against theoretical templates. When a "signal" fits, it is announced as the detection of spacetime ripples from black hole collisions billions of light-years away.

In both cases, the interpretation hinges on a symbolic system that only a select few understand. The public never sees the raw phenomenon directly — they see a *translated* version.

And as with the ancient priests, the authority lies in the confidence of the interpreter, not in the layperson's ability to verify the process.

The risk, then as now, is that when the symbolic system drifts away from physical reality, the predictions may still "work" inside the system — but they no longer correspond to what is actually happening. The priest may blame an angry god; the physicist may adjust a constant. The pattern remains: the guardians of the symbols control the narrative.

1.3 The Seduction of Mathematical Perfection

Humans crave certainty, and mathematics offers it in abundance. Numbers do not stutter. Equations do not contradict themselves. A calculation carried out correctly will yield the same result every time, regardless of who performs it. This precision is intoxicating in a world where reality is messy, unpredictable, and often resistant to neat explanations.

Mathematics also seduces through elegance. An equation that compresses a vast phenomenon into a single line of symbols is treated almost as a work of art. Scientists speak of "beautiful" theories with the same reverence as poets speak of perfect verses. Internal consistency — the ability of a system to obey its own rules flawlessly — gives it an air of inevitability. If the numbers line up and the logic holds, we instinctively feel that the theory *must* be true.

Over time, this led to a powerful cultural mantra in science: If the math works, the theory is valid. It is the modern equivalent of "the oracle has spoken." Equations have become more than tools; they are verdicts. Few stop to ask whether the model's starting assumptions are anchored to reality — or if the symbols merely dance in a closed loop of their own invention.

The danger is that mathematics can be perfectly self-consistent while being entirely disconnected from physical cause-effect. The internal beauty of a theory does not guarantee its external truth.

Consider a few celebrated examples:

- **Singularities** – In general relativity, the equations describing a black hole's core yield infinite density and zero volume. This "infinity" is not a measurable thing in nature — it is a mathematical breakdown. Yet it is often discussed as if it exists physically.
- **Multiverse Theories** – Certain solutions in string theory or quantum cosmology predict countless universes, each with different physical laws. These arise not from observation but from the way the math behaves under specific assumptions. No ripple-field evidence exists; they are pure extrapolations.
- **Cosmic Inflation** – To explain the uniformity of the cosmic microwave background, physicists introduced a phase of faster-than-light expansion immediately after the Big Bang. It fits the equations beautifully — but is unobservable and violates physical cause-effect as understood in UED.

Each of these is a triumph of mathematical imagination, but imagination nonetheless. When the elegance of the equation overshadows the grounding of the model, science quietly drifts into myth-making. The map, no matter how flawless, starts to replace the territory.

In **Universal Energy Dynamics**, mathematics is restored to its rightful place: a servant of measured cause-effect, never its master. An equation that fits the data but fails to align with ripple-field mechanics is not truth — it is a story in symbols, waiting to be tested against the real universe.

1.4 UED's Reality-First Filter

Universal Energy Dynamics (UED) begins with a simple but uncompromising principle: Mathematics is a map, not the territory.

Maps can be precise, elegant, and incredibly useful — but they are still only representations. A map can contain every contour line and every road, yet it will never *be* the mountain or the city it depicts. In the same way, equations can model reality with exquisite detail, but they are still abstractions. They must be judged by how accurately they reflect physical cause-effect, not by their internal beauty.

In UED, validation does not come from equation elegance or the ability to produce a "working" solution inside a closed symbolic system. Reality is confirmed only when the model aligns with two core criteria:

1. **Ripple-field causality** – Every event in the universe emerges from the interaction of structured vacuum energy (SVE) through ripple alignments. If a theory does not fit into this cause-effect chain, it may be mathematically correct but physically irrelevant.

2. **ATT resolution (Alignment Threshold Time)** – Observations are not instantaneous; they occur when alignment meets and consequences resolve. A model must account for this delay, recognizing that some "effects" recorded by instruments may be perception events, not real-time physical changes.

This shift reclaims mathematics from the throne it has quietly assumed in modern science. Under UED, math returns to being a servant of reality — a powerful tool for describing, predicting, and refining our understanding of the ripple field — but never the final judge of truth. The equations follow the physics, not the other way around.

This Reality-First Filter will be our lens for the rest of this book. In the chapters ahead, we will pass some of modern science's most celebrated ideas through it — from time travel and singularities to multiverse speculation and cosmic inflation. Many will crumble when stripped of their mathematical mythology. Others may survive, but in a humbler, more physically grounded form.

By the end of this journey, you will see why UED insists that the universe does not bow to the blackboard — and why, when theory and reality disagree, it is always the theory that must change.

Chapter 2 – The Mythology Template

Picture this: An ancient storyteller leans forward, eyes glinting in the firelight. "Every morning," he says, "the sun god mounts his golden chariot and rides across the sky. At night, he descends into the underworld, battling demons until dawn." The children stare, captivated. The story feels complete. It explains the sunrise and sunset perfectly — inside its own rules.

Now, picture this: A physicist stands before a whiteboard, eyes glinting with the same intensity. "At near-light speed," he says, "time dilates. A traveler could leap centuries into the future without aging. It's all here in the equations." The graduate students stare, captivated. The theory feels complete. It explains time travel perfectly — inside its own rules.

Two worlds. Two languages. One template. Both are *self-contained systems* with their own internal logic. Both use a *symbolic language* — one with gods and chariots, the other with constants and tensors. Both create a vision of reality that feels satisfying, even if no one can step inside to see if it's really there.

In this chapter, we'll strip myth and math to their bones to reveal the same underlying skeleton — the *mythology template*

— and see how modern science sometimes builds its grandest castles on the very same foundation as ancient legends.

2.1 The Skeleton of a Story

When we think of "myth," we often imagine ancient tales filled with gods, monsters, and heroic journeys. But a myth is not defined by its characters or setting — it's defined by its structure. Whether told by a poet, a priest, or a physicist, a myth follows a recognizable blueprint that makes it both convincing and enduring.

At its core, the structure of a myth has three universal elements:

1. Self-Contained Consistency

A good myth is airtight within its own logic. The rules of the story never contradict themselves. In the Greek myth of Persephone, her annual descent to the underworld explains the changing seasons; nothing in the story violates that premise. In physics, a self-contained model might explain time dilation or black hole evaporation perfectly — as long as you stay inside the assumptions of the model. This internal harmony gives the illusion of truth, even if the system has no anchor in reality.

2. Symbolic Language

Every myth speaks in symbols. In ancient epics, these are images and metaphors — golden chariots for the sun, serpents for chaos, rivers for time. In modern science, the symbols are equations, constants, and variables. They look objective, but to the untrained, they are just as mysterious as ancient runes. Only the "initiates" — whether temple priests or theoretical physicists — can read and interpret them, giving them authority over the narrative.

3. Appeal Without Proof

A myth doesn't need evidence to be persuasive. It needs emotional or intellectual resonance. Myths satisfy curiosity by giving complete answers, even when the answers are untestable. In the same way, a mathematical theory that predicts exotic realities — wormholes, multiverses, singularities — can satisfy intellectual curiosity without offering direct physical verification. The satisfaction comes from the elegance of the explanation, not from empirical confirmation.

The Double Edge

These traits are not inherently bad. In fact, they are part of how humans think and communicate. A consistent, symbolic, appealing model can be a powerful tool for exploring ideas — as long as we remember it is a model. The danger begins when we forget this distinction, when we mistake the map for the territory, the equation for the event, the story for reality.

It is here that mythology and certain strands of modern theoretical physics share common ground. Both can create fully functional worlds that seem real — until we try to step inside and find that the door doesn't exist.

2.2 Myths in Words, Myths in Numbers

Strip away the cultural layers and you'll find that myths — whether ancient or modern — are just **stories told in code**. The code changes, but the effect remains the same: it paints a picture of reality that feels vivid, coherent, and persuasive, whether or not it's physically true.

Consider these two statements:

- **Ancient Myth:** *"The sun rides across the sky in a golden chariot."* To the listener, this is a complete explanation for the daily movement of the sun. The chariot is a metaphor, the gold a symbol of divinity, the journey a reflection of cosmic order. The story is simple, memorable, and internally consistent.
- **Modern Math Myth:** *"Time curves and carries you into the future at near-light speed."*
  To the listener, this too feels like a complete explanation. The "curvature" is a metaphor, the constants and variables stand in for the laws of nature, and the image of a traveler stepping out into a far future satisfies the same craving for wonder and closure.

At first glance, these statements seem worlds apart — one belongs to an ancient storyteller by a fire, the other to a physicist in a lecture hall. But they share the same narrative mechanics:

1. **Symbolism shifts form** — The gods and chariots of myth become ccc, GGG, and tensor fields. In ancient stories, a divine being moves the heavens; in modern ones, an abstract variable manipulates spacetime. The symbols differ, but both are tokens of authority in their respective worlds.
2. **The authority delivers the code** — In a village, few could read the sacred scripts; they relied on the priest

or scribe. Today, few can follow the tensor calculus; they rely on the scientist or science communicator. In both cases, the public accepts the explanation largely because it is spoken by someone with access to the "language of truth."

3. **The public's role is the same** — Whether listening to a mythic tale or a physics lecture, most people are not evaluating the claim through direct experience. They cannot climb into the heavens to see the chariot, nor can they travel at near-light speed to test time dilation firsthand. They must trust the interpreter.

The problem isn't that symbolism exists — it's that we often forget it *is* symbolism. When we take it literally, the golden chariot becomes as "real" as the curvature of time, even though both may be narrative devices rather than physical truths.

2.3 Building Worlds Without a Door

One of the most captivating powers of both mythology and ungrounded mathematical models is their ability to create worlds that are so vividly imagined, so precisely described, that they feel real — even when no one can actually enter them. These are worlds without doors: places you can picture in detail, navigate in your mind, even map out on paper, yet never physically visit or measure.

In ancient mythology, these unreachable worlds were often moral or cosmic in nature.

- **The Mythic Underworld** — Described with stunning detail in

Greek, Egyptian, and Mesopotamian traditions, the underworld had gates, guardians, rivers, and cities. Poets could chart its geography as if they had been there. But it was accessible only through death, which meant no living person could test the description. The consistency of the story made it *feel* plausible, even though it was forever beyond direct verification.

In modern physics, the same narrative architecture appears — only the symbols change.

- **The Black Hole's Singularity** — According to general relativity, the center of a black hole collapses to infinite density and zero volume. It is described mathematically with exquisite precision. Researchers publish detailed predictions of what happens "inside." But no instrument can survive the trip to confirm it. Like the underworld, the singularity is beyond all possible observation — a place we can imagine perfectly but never reach.

Ancient mythmakers also offered visions of heavenly realms: paradises above the sky, layered with light and song, accessible only to the worthy after death. These were rich in imagery but closed to living eyes. Modern theory offers its own version:

- **Higher-Dimensional Spaces in String Theory** — These are precise mathematical constructs, often 10 or 11 dimensions, curled up beyond our perception. Entire frameworks of particle behavior are explained through these hidden realms. But they cannot be opened or walked into; they exist purely as requirements of the equations.

The psychological pull is easy to understand. Humans are drawn to complete, closed systems — worlds where every detail has an explanation and no contradictions remain. Such systems give us the pleasure of certainty and the comfort of order. They satisfy the intellect the way myths satisfy the heart. Whether the gates are guarded by Cerberus or hidden behind Planck-scale geometry, the effect is the same: a compelling illusion of reality without physical access.

The danger comes when we confuse mental completeness with physical truth. Just because a world is consistent and describable does not mean it exists. Without a door — without a way to align with it in the ripple field and cross the **Alignment Threshold Time (ATT)** — it remains a story, no matter how intricate the telling.

2.4 The UED Reality Access Test

Universal Energy Dynamics offers a simple but decisive way to separate physical reality from elegant fiction: the Reality Access Test. At its heart is the recognition that if something cannot be aligned with the ripple field and crossed through the Alignment Threshold Time (ATT), it remains a narrative — no matter how convincing the story or how beautiful the mathematics.

The Door Test

Think of ATT as the *door* between imagination and measurable reality.

- If an idea describes a world we can align with through ripple-field causality — meaning there's a measurable, physical sequence from cause to effect that survives scrutiny — then it's a candidate for real physics.

- If that world can be crossed into by reaching its ATT — meaning there is a minimum alignment period after which a consequence can be directly perceived or measured — it has physical grounding.
- If it fails either condition, it's a closed room with no entrance: internally detailed, possibly awe-inspiring, but sealed off from reality.

This test applies equally to ancient myths and modern mathematical constructs.

- The mythic underworld fails because no living observer can align with it in the ripple field.
- The black hole singularity fails for the same reason — its predicted state lies forever beyond our ability to cross the ATT boundary.

Why This Matters

A theory can survive in textbooks and journals for decades on elegance alone. But elegance is not evidence. By insisting that every proposed "world" be both alignable and accessible, UED ensures that mathematics serves reality, not the other way around.

The Coming Reality Checks

In the chapters ahead, we'll apply the UED Reality Access Test to some of the most celebrated constructs in modern science:

- **Time travel** – Can we ever align with a "future" or "past" that hasn't or no longer exists?
- **Singularities** – Is "infinite density" a physical state or just a mathematical placeholder for "we don't know"?
- **Dark matter** – Is it a real substance in the ripple field or an invented patch for flawed models?
- **Quantum collapse** – Is this a literal event or a perception artifact caused by ATT resolution?
- **Cosmic inflation** – Did the early universe really outrun light, or is this another doorless world?

Each of these will be tested not by how well they work on paper, but by whether they have a measurable, causal path through the ripple field and across ATT. Many will not survive the journey. Those that do will emerge stripped of myth and redefined in terms of physical reality.

Because in UED, the question is not *Does the math work?* but *Can the door be opened?*

Chapter 3 – UED's Reality Filter

Imagine a lie detector that doesn't measure your pulse or your sweat, but the very structure of reality itself. A device that could listen to a theory, run it through the universe's own logic, and instantly tell you: *This is real* or *This is only a story.*

That's what **Universal Energy Dynamics** offers — not in the form of a machine, but as a way of thinking. A filter. A truth test.

Because here's the problem: in science today, a theory can survive — even thrive — for decades on the strength of its mathematics alone. If the equations are consistent and the predictions fit the narrative, the model is often accepted, even if no one can ever measure the thing it describes. In such cases, we're no longer doing physics; we're doing mathematical storytelling.

UED breaks that cycle by asking three unforgiving questions:

1. **Alignment** – Can the observer's field actually match the event's ripple configuration?
2. **ATT (Alignment Threshold Time)** – Can alignment stabilize long enough for the consequence to be perceived or measured?

3. **Ripple-Field Causality** – Can every step of the phenomenon be traced through the structured vacuum's ripple interactions?

If the answer to any of these is *no*, the theory doesn't just fail — it dissolves into narrative, no matter how beautiful the symbols on the blackboard.

In this chapter, we'll see how this Reality Filter works, why it's needed, and how it can tell the difference between a mathematically perfect myth and a physically grounded truth.

3.1 Why a Filter Is Needed

Science is supposed to be about reality. Yet, over the last century, some of its grandest ideas have drifted into a space where they are mathematically flawless but physically hollow. These theories look bulletproof on paper. They have equations that balance perfectly, predictions that seem internally consistent, and even the approval of prestigious institutions. But when you ask the most basic question — *Can we actually align with this in the real universe?* — the answer is either "not yet" or "never."

The problem is subtle. Mathematical elegance has become a kind of gold standard for truth. A theory with neat, symmetrical equations feels right in the same way a perfectly rhymed poem feels complete. Internal consistency — the absence of contradictions within the system — adds to that feeling. When both are present, it's easy to mistake beauty for reality.

But elegance and internal logic can hide a fatal flaw: the absence of measurable cause-effect. If the steps of a theory cannot be traced through an actual physical process — if no ripple-field alignment can occur, or if the Alignment Threshold Time (ATT) for observation is unreachable — then the theory, no matter how "perfect," is a closed loop of symbols. It describes a world that exists only in its own language.

The risk is that science, built on such symbolic self-consistency, becomes detached from its grounding. It starts to resemble the very myths it once claimed to replace — not because it's wrong in intention, but because it confuses internal truth (truth within the model) with external truth (truth in the universe). Without a filter to separate the two, the blackboard becomes an altar, and we begin to worship the equations themselves.

This is why **UED's Reality Filter** is necessary. It forces every theory to step outside the comfort of its own logic and face the question: *Can it exist in the ripple-field reality we actually inhabit?* If not, then no matter how elegant, it belongs to the realm of narrative — not nature.

3.2 The Three Pillars of the UED Truth Test

The UED Reality Filter is built on three non-negotiable pillars. If a theory or claim fails even one of them, it cannot be called physical reality under Universal Energy Dynamics. These pillars are not abstract ideals — they are operational conditions that reality itself demands.

1. Alignment – The Gateway to Reality

In UED, *reality begins* when the observer's field configuration matches the event's ripple pattern.

- Every phenomenon — from a sound wave to a galaxy collision — manifests as a unique ripple arrangement in the structured vacuum energy (SVE).
- For an observer to register it, their local field must *align* with that pattern. Without alignment, the event might still exist, but it is effectively invisible and unreachable.
- Alignment is not a metaphor for "understanding" — it is a physical condition, like tuning a radio to the right frequency before you hear the broadcast.

Example: A radio telescope can detect signals from a distant quasar only if its receiving system is configured for the right wavelength. In UED terms, the telescope+observer system must align its ripple field to that of the quasar's emission before any data is real in the physical sense.

2. ATT (Alignment Threshold Time) – The Door Must Stay Open Long Enough

Alignment alone isn't enough — it must stabilize for a minimum duration before any consequence can be perceived or measured. This is the Alignment Threshold Time (ATT).

- ATT represents the *minimum time the "door" to an event must remain open* for a detectable consequence to occur.
- If the alignment is broken before ATT is reached, the phenomenon leaves no measurable trace, even if the

math predicts it should.

- Many supposed "discoveries" in physics are actually *pre-ATT perception artifacts* — interpretations of patterns detected before stable alignment, which may never correspond to a physical event.

Example: In mainstream time-travel claims, the "arrival" in the future is just a reconfiguration of perception before the ATT of that future's ripple field could ever exist. You cannot measure what hasn't crossed its ATT; it's not there yet.

3. Ripple-Field Causality – The Chain Must Be Unbroken

Every valid physical event must be traceable through a coherent chain of ripple interactions in the SVE.

- This chain is the cause-effect path — the real physical mechanism by which the event happens.
- If any part of the chain is missing, undefined, or requires "magic constants" to work, the theory is narrative, not physics.
- Ripple-field causality ensures that no phenomenon is allowed to "float" on paper without physical grounding.

Example: In UED, gravitational effects aren't caused by "mass curving spacetime" — they are the result of ripple tension patterns in the SVE. Every effect has a physical ripple-chain explanation; nothing is left hanging in symbolic space.

How the Three Pillars Work Together

A phenomenon passes the UED Truth Test only if:

1. **Alignment** occurs — the observer can match the event's ripple pattern.
2. **ATT** is crossed — the alignment lasts long enough for consequences to register.
3. **Ripple-field causality** is intact — the event has a continuous, measurable cause-effect chain in the SVE.

If all three are met, the event belongs to the physical universe. If one fails, the event exists only in the realm of mathematical or narrative imagination — regardless of how elegant the theory is or how famous its proponents are.

This triad is the heart of UED's Reality Filter, and in the coming sections, we'll see it applied like a scientific lie detector to the most celebrated "truths" of modern science.

3.3 The Filter in Action

The **UED Reality Filter** is not a philosophical exercise — it is a working test you can apply to any proposed phenomenon, whether ancient myth or cutting-edge theory. The process is brutally simple: check for alignment, ensure ATT can be crossed, and trace ripple-field causality without gaps. If the proposal fails at any step, it's not physics — it's narrative.

Validated Examples – Theories That Pass the Test

Planetary Orbits

- **Alignment:** The Earth's field configuration continually aligns with the Sun's gravitational ripple pattern.

- **ATT:** The alignment is stable over billions of years — consequences are visible and measurable continuously.

- **Ripple-Field Causality:** Orbital motion is a ripple tension equilibrium in the SVE, fully traceable through observed and calculable cause-effect.

Verdict: Passes. Reality we can live in and measure directly.

Sound Propagation

- **Alignment:** The air's ripple pattern changes with the vibration source, and our ears (or instruments) align to receive these changes.
- **ATT:** Stable over microseconds to seconds, enough for perception or recording.
- **Ripple-Field Causality:** Chain from vibrating object → medium ripple → detection is unbroken.

Verdict: Passes. Everyday, directly testable physics.

Material Stress Limits

- **Alignment:** Load application and structural ripple-field pattern match during testing.
- **ATT:** Sufficient stability to measure strain before failure.
- **Ripple-Field Causality:** Deformation arises from traceable ripple-chain changes in atomic/molecular

alignment.

Verdict: Passes. The results are consistent and repeatable in any lab.

Rejected Examples – Theories That Fail the Test
Time Travel Loops

- **Alignment:** Cannot occur with a ripple pattern that hasn't yet formed (future) or has already dissipated (past).
- **ATT:** Impossible to cross because the "destination" never reaches stable alignment with the traveler's field.
- **Ripple-Field Causality:** Broken chain — no physical mechanism connects traveler to an unformed or erased configuration.

Verdict: Fails. Pure narrative, sustained by perception-based math.

Infinite Singularities

- **Alignment:** No observer can align with an infinite-density state; it's a mathematical placeholder, not a physical pattern.
- **ATT:** Cannot be crossed — the singularity is beyond the event horizon's causal reach.
- **Ripple-Field Causality:** Chain terminates at an undefined condition ("infinity"), which is not a cause-effect step.

Verdict: Fails. Doorless world.

Unmeasurable Extra Dimensions

- **Alignment:** No physical alignment possible with a dimension curled beyond detectability.
- **ATT:** No measurable interaction to stabilize.
- **Ripple-Field Causality:** No chain — dimensions are introduced to satisfy math, not observed reality.

Verdict: Fails. Symbolic speculation.

Why the Filter Doesn't Kill Imagination

Some might fear that applying such a strict filter would strangle creativity in science. In reality, the opposite is true. The UED filter channels imagination into reachable worlds. It doesn't say, "Don't imagine." It says, "If you imagine, give your world a door we can open."

Great science is not about trimming our curiosity — it's about ensuring that curiosity leads us somewhere we can actually go. In this way, the UED Reality Filter is not a wall; it's a funnel that directs the boundless energy of human speculation into pathways that intersect with the real universe.

3.4 From Myth to Measurable

The UED Reality Filter is not designed to simply discard ideas that fail. Its deeper value lies in transforming them — reshaping speculative constructs so they can pass into the realm of testable science. In other words, it turns myth into measurable.

When a theory fails one of the three pillars — alignment, ATT, or ripple-field causality — it doesn't have to be abandoned. Instead, it can be re-engineered:

- If alignment is impossible, the model can be reframed so that the event's ripple pattern becomes physically matchable by an observer or instrument.
- If ATT cannot be crossed, the design of the experiment or the definition of the phenomenon can be adjusted to allow stable alignment long enough for measurement.
- If ripple-field causality is broken, missing steps can be identified, replaced, or reinterpreted within the structured vacuum energy (SVE) framework until the chain is complete.

This process keeps creativity alive while tethering it to the physical universe. Instead of chasing mathematical perfection in a vacuum, researchers are encouraged to give their ideas a real-world doorway.

Mathematics After the Filter

Once a theory passes the Reality Filter, mathematics returns to its rightful place:

- **Refining** – describing the phenomenon with precision so experiments can be tightly designed.
- **Predicting** – extending the model to forecast behavior in untested scenarios, which can then be verified.
- **Scaling** – applying the phenomenon to larger or

smaller domains without breaking the cause-effect chain.

But mathematics never dictates reality. In UED, math is a measuring stick, not a ruler. It measures the territory; it does not define it.

A Foundation for What Comes Next

This filtering framework sets the stage for the myth-busting journey ahead. In the following chapters, we will take celebrated ideas — time travel, singularities, dark matter, quantum collapse, cosmic inflation, and more — and run them through the UED Reality Filter.

Some will fall apart instantly when the door test is applied. Others will survive but emerge looking nothing like their original form, stripped of their mythical layers and rebuilt as physically grounded hypotheses.

The point is not to belittle these ideas but to liberate them from the prison of symbolic self-consistency — to return them to a reality where cause and effect can be touched, measured, and understood.

Because the ultimate test of any theory is not whether it works on the blackboard, but whether it can be opened like a door and stepped into by reality itself.

Interlude – Stepping Back from the Blackboard

We have arrived at a turning point. In the past three chapters, we have pulled the curtain back on a quiet but powerful truth: mathematics, when unanchored from physical cause-effect, becomes a form of storytelling. It can be as vivid, intricate, and internally consistent as any ancient myth — and just as unreachable.

We began with **Chapter 1**, tracing the evolution of mathematics from a humble counting tool to a seat of unquestionable authority. We saw how numbers, once servants of reality, came to be treated as its masters — much like the priests of old who claimed to speak for the gods.

In **Chapter 2**, we stripped both myth and math down to their bare bones and discovered the same skeleton — self-contained consistency, symbolic language, and an appeal that doesn't require proof. The language changes, but the structure is timeless. The storyteller may now wear a lab coat instead of ceremonial robes, yet the narrative mechanics remain identical.

In **Chapter 3**, we introduced **UED's Reality Filter** — a simple, uncompromising test to separate elegant fiction from grounded physics. Alignment, **Alignment Threshold Time**

(ATT), and ripple-field causality became our three pillars. These are not just abstract ideas; they are the conditions reality itself demands before something can be called real.

This is more than an academic exercise. It is a reset. We are no longer willing to accept the blackboard's authority without question. From here on, every idea — no matter how celebrated, how "proven" in its own symbolic world — must step through the UED door test. If it cannot be aligned, crossed through ATT, and traced through ripple-field causality, it will be revealed for what it is: narrative, not nature.

In **Part II**, we will put this filter to work. Theories that have dominated headlines, captivated imaginations, and filled lecture halls will be tested without mercy. Some will survive, transformed. Others will crumble under the weight of their own symbolism. And in the process, we'll discover something unexpected: stripping away the myth doesn't make science dull — it makes reality more astonishing than the fiction ever was.

Part II – Case Studies in Modern Mathematical Myths

Chapter 4 – Time Travel and the River That Never Was

We've all heard the story: time is a great river, flowing endlessly from the past into the future. If you could build the right boat — a spaceship near the speed of light, perhaps — you could sail ahead into tomorrow or drift back into yesterday. It's a poetic image, comforting in its clarity, and endlessly recycled in science fiction and late-night debates.

Mainstream physics, through relativity, seems to offer this river as more than a metaphor. It presents time as a fourth dimension, a coordinate you can move through like length, width, or height. Bend it with gravity, stretch it with velocity, and suddenly, "time travel" becomes not just fantasy, but mathematics. The idea is intoxicating: step into your machine, and step out a hundred years from now — or a hundred years ago.

But here's the truth under **Universal Energy Dynamics (UED)**: there is no river. There is no "ahead" or "behind" in time. What we call "time" is not a dimension waiting to be navigated, but a delay — the **Alignment Delay** between when a ripple-field configuration aligns and when its consequences resolve. The so-called river is a mirage created by perception before the **Alignment Threshold Time (ATT)** is crossed.

And when you remove the river, all the paradoxes — the loops, the contradictions, the impossible meetings of cause and effect — vanish instantly. The grandfather can rest easy, because you were never going to meet him before you were born.

In this chapter, we'll drain the river entirely and see what remains: a universe where "time travel" collapses, but reality becomes clearer — and stranger — than any science fiction ever dared to imagine.

4.1 The Mainstream Story of the River

In the early 20th century, Albert Einstein's theories of special and general relativity reshaped the way physics described the universe. Time was no longer treated as a separate, universal clock ticking the same for everyone. Instead, it became part of a four-dimensional entity called spacetime — a seamless fabric where three dimensions of space and one of time are woven together.

In this framework, time behaves like a coordinate, similar to distance. Move through space in one way, and you move through time in another. According to the equations, both velocity and gravity can "bend" or "stretch" time:

- At high speeds, a moving clock runs slower compared to one at rest — the effect known as time dilation.
- In strong gravitational fields, clocks tick more slowly relative to those in weaker fields — a phenomenon called gravitational time dilation.

From this starting point, the story becomes irresistibly romantic. If moving fast or sitting deep in a gravity well slows your passage through time, then perhaps you could use this to leap into the future. The concept of "time travel" gains an aura of scientific legitimacy:

- **Future travel**: Fly a spaceship at near-light speed for what feels like a few years to you, and return to find decades or centuries have passed on Earth.
- **Past travel**: Twist spacetime into loops, perhaps through wormholes — hypothetical tunnels connecting two points in spacetime — and step into an earlier era.

The equations allow for both scenarios under certain interpretations, and that's all the opening needed for science fiction — and even some scientists — to embrace the vision of a navigable "river of time."

Popular culture amplified the imagery until it felt as real as gravity itself. Movies, novels, and television dramas have filled the public imagination with time machines, paradoxes, and altered histories. From *H.G. Wells* to *Interstellar*, the river of time became a shared cultural myth — one that seemed to have mathematics on its side.

But in the excitement, a crucial assumption was rarely questioned: does time truly flow like a river, waiting for us to enter and travel along? Or is this river nothing more than a mirage born from perception and the limitations of measurement?

That's the question UED will now confront.

4.2 The Birth of the Paradox

Once you accept the idea of time as a flowing river — something you can move through as freely as you cross a stream — paradoxes emerge almost immediately. The most famous is the grandfather paradox:

Imagine you travel back in time and prevent your grandfather from meeting your grandmother. Without their union, your parent would never be born, and therefore, neither would you. But if you were never born, how could you have gone back in time to stop their meeting in the first place? The chain of events folds back on itself, creating an impossible loop.

Physics gives this problem a formal name: the causal loop — a sequence of events where cause and effect become tangled to the point of logical impossibility. And once the river model is accepted, these loops are unavoidable:

- If the "past" still exists and can be accessed, then the events that made the present can be altered.
- If the "future" already exists and can be visited, then actions in the present might be influenced by something that hasn't "happened" yet.

Mainstream models tolerate these paradoxes because their mathematics treats time as a continuous, navigable dimension. In that framework, every point in time is like a location on a map: it's already there, waiting for you to visit. Under this view, the only question is how to get from one point to another — not whether those points physically exist in the first place.

The fatal flaw is subtle but decisive: it conflates changes in perception before full physical realization with actual changes in time itself.

- At high velocities or in strong gravity, the rate at which processes appear to unfold for different observers can diverge. This is a change in perception — the way events are *registered* — not a rewinding or fast-forwarding of a universal clock.
- The "future" you think you're traveling to in these scenarios doesn't yet exist in the ripple field. The "past" you think you can revisit has already dissipated. In both cases, there is no stable configuration to align with, and without alignment, there is no physical access.

In other words, the paradox is not a property of the universe — it's a property of a mistaken model. Remove the river, and the loops vanish instantly. The universe is not tangled; only our storytelling about it is.

4.3 UED's Redefinition of Time

In **Universal Energy Dynamics (UED)**, time is stripped of its mystique and treated not as a flowing dimension but as a simple, measurable delay. This delay — the gap between a ripple-field alignment and the full resolution of its consequences — is what we perceive as time.

Time as Alignment Delay

Every event in the universe begins when the observer's field configuration aligns with the event's ripple pattern in the structured vacuum energy (SVE). But alignment is not the same as instant perception. There is always an Alignment Delay — the interval between the moment alignment occurs and the moment the event's consequences become fully observable or measurable.

- This delay is not arbitrary; it is determined by the physics of ripple interaction, distance, medium, and field stability.
- What we call "time passing" is nothing more than this sequence of alignments and consequence resolutions happening continuously in our awareness.

ATT as the Gatekeeper

The **Alignment Threshold Time (ATT)** is the minimum duration for which alignment must remain stable for an event to become part of your experienced reality.

- If alignment is broken before the ATT is reached, the event never becomes physically accessible to you — it remains unreal in your frame of reference.
- ATT is not a human construct like the second or the hour; it is a natural threshold dictated by ripple-field physics. It defines when a phenomenon is "real" for an observer, regardless of what the clock says.

In practical terms, ATT is the universe's *gatekeeper*. Only events that survive long enough through stable alignment to cross this threshold are "let in" to your reality. Everything else stays outside, like an unfinished ripple that never reaches shore.

No Future, No Past — Only Configurations

This model immediately dismantles the idea of traveling through time.

- The "future" does not exist as a formed landscape waiting to be reached. It is simply a set of ripple configurations that have not yet aligned or stabilized.
- The "past" is not a preserved archive that can be revisited. Once its ripple configurations have dissipated and reconfigured, they no longer exist in their original form anywhere in the SVE.

In UED, there is no river of time flowing past you. Instead, you are continually encountering and aligning with fresh configurations in the ripple field. When an event has not yet formed, there's nothing to travel to; when it has dissipated, there's nothing to go back to. The only reality is the configuration you are aligned with now, and the measurable delay until its consequences fully arrive.

This redefinition doesn't just make paradoxes impossible — it makes them meaningless. The concept of "looping" through time assumes a structure that UED shows does not exist in the first place.

4.4 Why Paradoxes Collapse in UED

In mainstream time travel models, paradoxes emerge because time is treated as a navigable path — a trail you can walk backward to the past or forward to the future. Once you picture time as a track already laid out, it's easy to imagine jumping to a different point on it. But UED removes the track entirely, and with it, every paradox that depends on its existence.

No Temporal Path to Retrace

In UED, there is no pre-existing timeline stretching behind you or ahead of you. Reality is not a reel of film where each frame is still "out there" somewhere, waiting to be replayed. Instead, it is a continuous unfolding of ripple-field configurations, each forming and dissolving in the structured vacuum energy (SVE). Once a configuration has changed, the previous one no longer exists anywhere — it has been reabsorbed into the field.

- You cannot "return" to a past configuration because there is no physical pathway back to it.
- You cannot "advance" to a future configuration because it hasn't formed or stabilized yet.

Without a temporal path, loops like the grandfather paradox are structurally impossible. There's nothing to loop through.

No Interaction Without Stabilization

Every physical interaction requires alignment with a ripple configuration and enough time for that alignment to cross the **Alignment Threshold Time (ATT)**. If a configuration hasn't stabilized, there is literally nothing there to interact with — no particles, no fields, no consequences to experience.

- The "future" you think you could visit is only a set

of potential ripple arrangements. Until they align and stabilize, they are unmanifested possibilities, not destinations.

- The "past" is not a frozen scene you can walk back into; its configurations are gone, replaced by the ripple patterns of the present.

Trying to interact with either is like trying to touch the reflection of a bridge that has not yet been built or has already been dismantled.

The Future Traveler's Illusion

In mainstream terms, a "future traveler" is someone who experiences time at a different rate — for example, due to high velocity or strong gravity — and then emerges into what others perceive as the future. In UED, this is simply a shift in perception within a pre-ATT window.

- The traveler's local field configuration experiences fewer perceptual updates than the surrounding environment.
- Upon returning to normal conditions, they align with a configuration that others reached earlier, creating the illusion of having "jumped forward" in time.
- No actual movement through a temporal dimension occurred — only a variation in how and when alignments were registered.

This is why the "river of time" analogy fails. You are not floating downstream faster or slower than others. You are sitting in your own ripple pocket, receiving updates at a different rate, until you step back into the wider field.

By removing the temporal path, insisting on ATT stabilization, and reframing "future travel" as perception shifts rather than dimensional leaps, UED makes causal loops and time travel paradoxes impossible. The universe doesn't bend to fit our narrative contradictions — it simply never allows them to form.

4.5 The Real Science Hidden Behind the Myth

It's important to note that when UED dismantles the "river of time," it is not dismissing the physical phenomena that inspired it. Relativistic effects — the measurable differences in how processes unfold under high velocity or strong gravity — are real. What UED challenges is the interpretation of these effects as literal travel through a temporal dimension.

When we strip away the imaginary timelines, what remains is a set of well-defined physical processes that can be understood entirely in ripple-field terms.

Field Reconfiguration Under High Velocity

In UED, moving at high speed relative to a given reference field does not "slow down" time — it changes the configuration of your local ripple field.

- As velocity increases, the way your field aligns with incoming ripple patterns changes.
- Some ripple interactions require longer to stabilize; others become harder to maintain alignment with.

- This affects the rate at which you register and respond to events, creating the measurable differences we call "time dilation" in mainstream physics.

Rather than imagining you "slipping into the future," UED describes this as operating in a modified ripple environment. Your clock, your heartbeat, and your thought processes are all tied to the stability of your local field alignment, which shifts under high motion.

Variation in ATT Due to Environmental and Motion-Induced Ripple Changes

The Alignment Threshold Time (ATT) is not fixed — it can change depending on the surrounding ripple conditions.

- In strong gravitational fields, ripple tension is altered, affecting how quickly or slowly alignments stabilize.
- At high velocity, motion-induced distortions in the local ripple pattern can lengthen or shorten ATT for certain interactions.

This variation explains why two observers in different conditions — one on Earth's surface, one orbiting in space, or one traveling at a fraction of light speed — will record different intervals between equivalent events. They're not on different "timelines"; they're operating with different ATT thresholds.

Studying the Effects Without Imaginary Timelines

UED shifts the question from *"Where did the traveler go in time?"* to *"How did their ripple-field alignment and ATT change under these conditions?"*

- Instead of mapping a journey along a temporal axis, experiments can map changes in ATT and alignment stability as functions of velocity, field strength, and environmental ripple characteristics.
- This removes paradoxes entirely, because there is no assumption of a pre-existing "future" or "past" being accessed — only configurations that are presently forming or have already dissolved.

By focusing on the ripple-field mechanics rather than fictional timelines, UED retains all the measurable science — the clocks that disagree, the signals that arrive shifted — while discarding the narrative baggage that turns these effects into time travel fantasies.

In the UED framework, the "mystery" of time dilation becomes simply a matter of field behavior, not a bending of an invisible river. The science remains intact; the myth is gone.

Chapter 5 – Black Holes and the Infinity Illusion

Imagine falling toward a black hole. The pull grows stronger with every heartbeat. Space seems to stretch, light bends around you, and the stars smear into strange arcs. You cross an invisible boundary — the event horizon — and from that moment on, the outside universe can no longer see you.

Mainstream physics tells us that if you could keep going, you would be crushed into a singularity — a point of infinite density and zero volume. All the mass of the star that collapsed, and anything else that ever fell in, supposedly compressed into a dimensionless nothing. It's the ultimate "end of physics" — a place where our laws of nature break down and infinity reigns.

But in **Universal Energy Dynamics (UED)**, that singularity is nothing more than a mathematical ghost. The infinity doesn't exist in reality — it exists in the equations, born from pushing the math beyond its physical limits. Just as dividing by zero on a calculator returns an error, not a new universe, the singularity is a signal that the model has hit a wall.

What's really at the heart of a black hole? Not an infinitely small point, but a finite, hyper-compressed state — the maximum ripple-field compression the structured vacuum energy (SVE) can sustain. The mystery isn't infinity; it's the extreme but *bounded* nature of that compression.

In this chapter, we'll strip away the drama of the bottomless pit and replace it with the measurable physics of ripple fields, ATT delays, and maximum compression. The result is still astonishing — but it's grounded in reality, not in the illusion of infinity.

5.1 The Mainstream Black Hole Narrative

In mainstream physics, the story of a black hole begins with general relativity. According to Einstein's field equations, if enough mass is packed into a small enough space, gravity overwhelms all other forces, collapsing matter into a point of zero volume and infinite density — the infamous singularity. At this point, the equations themselves stop working. Physicists describe it as a breakdown of known laws, a place where spacetime curvature becomes infinite and no theory can predict what happens inside.

Surrounding this singularity is the event horizon, the so-called "point of no return." This is the boundary at which the escape velocity equals the speed of light. Anything crossing it — matter, light, even information — is thought to be forever trapped. From the outside, we can never see beyond this horizon. We can only infer what lies inside from indirect signals, like gravitational waves or X-ray emissions from infalling material.

The prediction of infinity in this model does not come from direct observation; it comes from the mathematics of general relativity. When the mass term is pushed to extremes in the equations, the curvature term spikes toward infinity. This is a mathematical consequence of the model's structure, not a physical measurement. In reality, we cannot probe or verify the internal state of a black hole — the singularity exists only as the logical end-point of the equation.

Nevertheless, this has not stopped the singularity from becoming a powerful image in both science and popular culture. The idea of an infinitely deep, bottomless pit where the known universe collapses into "nothing" captures the imagination. Movies, documentaries, and even some textbooks depict matter falling endlessly into the abyss, vanishing from existence as it crosses the event horizon. The story is cinematic, dramatic, and — in the public mind — definitive.

But it rests entirely on the assumption that the math can be extrapolated without limit into regions we can never observe. The "infinity" is not seen, measured, or tested. It is accepted because the equations predict it, and because the narrative is compelling. As we'll see, **UED** approaches black holes very differently — not as portals to infinity, but as extreme yet finite configurations of the ripple field, shaped by the maximum compression the structured vacuum energy (SVE) can sustain.

5.2 The Singularity: A Mathematical Ghost

In mathematics, "infinity" is not a number and certainly not a physical object. It is a conceptual placeholder — a way of saying "this value grows without bound" or "this equation cannot be resolved within its current framework." When infinity appears in physics equations, it does not mean the universe has produced an actual infinite thing; it means the model has been pushed beyond its range of validity.

This is exactly what happens in the mainstream description of a black hole singularity. General relativity predicts that as a massive star collapses under its own gravity, the curvature of spacetime increases. At a certain point, if we keep following the math without modification, the radius of the collapsing matter shrinks toward zero while the density shoots toward infinity.

But this "infinity" is not a measurable state of matter or space. It is the mathematical equivalent of dividing by zero:

- If you try to calculate 1/0, your calculator won't give you a physical answer — it gives you an error or an undefined symbol.
- The problem is not that the real world has suddenly become undefined; it's that your chosen method of describing it has broken down.

In the case of black holes, the singularity is a ghost entity — a phantom born from extending an equation into a region where it no longer applies. It's like drawing a map of the Earth and insisting the lines continue to infinity because you ran out of paper. The "infinite" is an artefact of the representation, not the terrain.

The danger lies in mistaking this ghost for reality. Once the singularity is accepted as a real, physical point of infinite density, an entire narrative forms around it — of spacetime collapsing into "nothing," of information vanishing forever, of a region where physics itself dies. This is powerful imagery, but it's imagery nonetheless. It substitutes a mathematical breakdown for an actual physical state.

Universal Energy Dynamics (UED) refuses to accept infinities as physical. Instead, it asks: what happens in the ripple-field when matter is compressed to its maximum possible extent within the structured vacuum energy (SVE)? The answer, as we'll see, is not infinity — it is maximum compression, a finite but extreme condition with measurable consequences.

5.3 UED's Redefinition: Maximum Compression Without Infinity

In **Universal Energy Dynamics (UED)**, the collapse of matter under extreme gravity does not lead to a magical leap into infinity. Instead, it follows a clear physical limit — a maximum compression threshold dictated by the nature of the structured vacuum energy (SVE) itself.

The SVE is not an empty void; it is the foundational ripple medium in which all matter and energy configurations exist. Every particle, field, and interaction is a ripple arrangement within this structure. When gravitational collapse begins, matter is pushed into tighter and tighter ripple configurations. This continues only until ripple-field tension — the resistance of the SVE to further compression — reaches its natural ceiling.

The Maximum Compression Threshold

- Ripple-field tension behaves like the ultimate stress limit of a material. Just as steel can be compressed only so far before it resists further deformation, the SVE can only be compressed to a specific density before the field's internal dynamics push back.

- At this limit, no additional collapse can occur without breaking the fundamental structure of the field itself — something nature does not permit without catastrophic reconfiguration.
- This maximum compression is finite. It is extreme beyond anything we encounter elsewhere in the universe, but it is still a definable, measurable condition.

The Hyper-Compressed State

Beyond this threshold, the black hole's core exists not as a mathematical point but as a hyper-compressed, stable configuration of matter-energy:

- All mass is stored in the densest possible ripple arrangement.
- The field becomes so tightly packed that no conventional particle movement is possible; the configuration is essentially "locked."
- Energy still exists within, but it is trapped in a condition where its release would require an unfeasible ripple-field reconfiguration.

No Point Singularity

In UED, there is no point-like singularity of zero volume. The core has a size, however small, defined by the SVE's compression limit. This size depends on the total mass-energy involved but will always remain above the zero-volume fiction of the singularity.

- The radius of this core may be far smaller than the Schwarzschild radius (the event horizon), but it is never undefined or infinite.
- Because the configuration is finite, it can, in principle, be described physically — its properties are extreme but real, not ghostly placeholders.

In this model, a black hole is no longer a bottomless pit ending in a physical impossibility. It is a finite object with a knowable internal limit, surrounded by a zone of extreme ripple distortion. This redefinition doesn't make black holes less mysterious — it makes them more fascinating, because it replaces the fantasy of infinity with the challenge of understanding one of the most extreme, yet physically lawful, states in the universe.

5.4 Observational Illusions from Perception-Before-ATT

The extreme gravitational environment near a black hole does more than just bend light — it also alters how and when events can be aligned and perceived. In Universal Energy Dynamics (UED) terms, the intense ripple-field distortion in this region directly impacts Alignment Threshold Time (ATT), creating effects that observers often misinterpret as physical changes in the object itself.

Gravitational Ripple Distortion and ATT

In the structured vacuum energy (SVE), strong gravitational fields are regions of high ripple tension and curvature. Near a black hole's event horizon:

- The ripple patterns between the infalling object and the distant observer become increasingly misaligned.
- The time it takes for these patterns to stabilize into a perceivable event — the ATT — stretches dramatically.
- This delay is not just a slow-down in perception; it is a real limitation in how quickly alignment can occur across such extreme field conditions.

As ATT increases toward infinity from the perspective of the distant observer, the moment of full alignment with the infalling object never arrives. The event becomes permanently incomplete from that vantage point.

Why Matter Appears to "Freeze" or Fade

Mainstream relativity predicts that an outside observer will never see an object actually cross the event horizon. Instead, it will appear to slow down, redshift, and eventually fade away. In UED, this is not because the object is "frozen" in space-time — it's because the observer's field can no longer achieve alignment with the object's ripple configuration beyond a certain point.

- The object continues to move toward the black hole in its own local frame.
- But for the distant observer, the alignment chain is stretched beyond measurable limits — the required ATT is never met.

- As a result, the object's final moments seem suspended forever, and its image fades as the signal's wavelength is stretched beyond detection.

The Perception Trap

Mainstream interpretations take this visual "freezing" as a literal representation of the object's physical state. But in UED, it is purely a perception-before-ATT effect:

- What is "seen" is not the present reality of the object, but the last fully aligned configuration transmitted before alignment became impossible.
- This means we are observing historical light — a ripple record — not the ongoing state of the object itself.

By failing to distinguish between what is aligned and observable versus what is physically occurring, mainstream narratives conflate perception delays with actual physical phenomena. The result is the illusion that matter somehow "hovers" at the horizon or vanishes in place, when in fact, it is simply beyond the reach of the observer's field alignment.

In UED, black holes remain mysterious, but they no longer require impossible physics or infinite time to explain what we see. The mystery lies in the extreme ripple dynamics and the boundaries of alignment — not in a frozen moment suspended forever in a mythical river of time.

5.5 The Real Black Hole in UED Terms

In Universal Energy Dynamics (UED), a black hole is not a cosmic magic trick or a mathematical singularity. It is a finite, physically definable object — a hyper-compressed core wrapped in one of the most extreme ripple environments in the universe.

The Finite, Hyper-Compressed Core

At the heart of the black hole lies matter-energy in its densest possible state, determined by the maximum compression threshold of the structured vacuum energy (SVE).

- The core is not infinitely small; it has a measurable, though incredibly tiny, radius set by the ripple-field's ultimate resistance to further compression.
- All mass-energy inside exists in a locked ripple configuration, with no conventional motion or particle structure — a stable, extreme equilibrium.
- This configuration is not a "point" but a volume, however small, that could be physically described in terms of ripple density and tension.

The Surrounding Region of Extreme Ripple Distortion

Beyond the core lies a region where ripple patterns are so intensely curved and tensioned that:

- Alignment between the black hole's field and a distant observer becomes increasingly difficult.
- Alignment Threshold Time (ATT) stretches toward unmeasurable limits, causing the well-known

observational illusions near the event horizon.

- Light and matter signals emitted from near the core are heavily redshifted and distorted, not because they are "lost," but because their ripple transmission fails to meet alignment requirements at distant locations.

This surrounding region — often confused with the "surface" of the black hole — is a transition zone, not a physical wall. It marks the boundary between accessible and inaccessible ripple configurations for a given observer.

No Infinities, No Ghost Points

In UED, the black hole contains nothing infinite, undefined, or "outside" the laws of physics:

- The core is finite and describable.
- The event horizon is not a mystical barrier, but a practical limit of alignment.
- The concept of a singularity is replaced by a physically lawful, measurable compression state.

This view removes the need for imaginary constructs like "information lost to infinity" or "points where physics ceases to exist." Instead, the black hole is a continuous physical system, fully integrated into the same ripple-field principles that govern the rest of the universe.

Implications for Mergers, Radiation, and Information Preservation

- **Black Hole Mergers** – In UED, when two black holes merge, their cores do not collide as point singularities; their hyper-compressed ripple configurations merge into a new, stable configuration. The gravitational wave signals we detect are ripple-field rebalancing events, not spacetime "shocks."

- **Radiation** – Phenomena like Hawking radiation can be reinterpreted as ripple leakage and reconfiguration at the edge of the alignment zone, not as particles "escaping" from an infinite well.
- **Information Preservation** – Since the core is finite and physically describable, information is not "destroyed" or lost to a mathematical infinity. It is transformed into a locked ripple pattern that remains part of the universe's total energy configuration, even if it is inaccessible to outside observers.

Under UED, the black hole is still one of the most extreme objects in the cosmos — but it is not a place where reality collapses into the impossible. It is a natural, finite endpoint of gravitational compression, obeying the same ripple-field laws as everything else, with mysteries that can be explored without invoking infinities or ghost points.

Chapter 6 – Dark Matter and Dark Energy

The Invisible Patchwork

If you believe the current cosmological story, almost everything you see — every star, planet, and glowing galaxy — is just the tip of the iceberg. According to mainstream models, over 95% of the universe is invisible. Not just dim, not just hidden from our telescopes, but made of stuff we have never directly detected: dark matter to hold galaxies together, and dark energy to push the cosmos apart.

It sounds like a shocking discovery. In reality, it was a bookkeeping fix. The math didn't match the measurements — galaxy rotation speeds were too high, cosmic expansion too fast — so new, unseen components were added to make the equations balance. First came the invisible glue (dark matter), then the invisible accelerator (dark energy). The story was written backward: equations first, mystery particles later.

Decades have passed, countless experiments have searched, and not a single confirmed particle of dark matter has been found. Dark energy is even more elusive — not a particle, not a field we can measure, but a placeholder for an effect we don't yet understand.

From **Universal Energy Dynamics (UED)**, this isn't a case of missing ingredients. It's a case of misreading the recipe. What we're seeing is not evidence of hidden matter and energy, but the large-scale behavior of ripple tension in the structured vacuum energy (SVE) — a real, physical mechanism that can produce the exact patterns we observe without inventing new invisible substances.

In this chapter, we'll peel back the layers of this invisible patchwork, show how the "fix-by-addition" approach turned cosmology into a modern myth factory, and see how UED's ripple-field model can make the universe whole again — without adding anything that isn't already there.

6.1 The Birth of the Invisible Components

The story of dark matter and dark energy begins not with a discovery, but with a discrepancy. In the early 20th century, astronomers began measuring the rotation speeds of galaxies. According to Newtonian gravity — and later, general relativity — stars farther from a galaxy's center should orbit more slowly, just as planets in the solar system move slower the farther they are from the Sun.

But that's not what the data showed. In galaxy after galaxy, the outer stars were moving just as fast as — or faster than — the inner ones. By the rules of the established gravity models, these galaxies should have flown apart. Something, it seemed, was providing extra gravitational pull to keep them intact.

Instead of questioning whether gravity itself might work differently on cosmic scales, mainstream cosmology took an "equations first" approach: keep the existing gravitational framework intact, and add an unseen component to make the math match the observation. That invisible mass was named dark matter — a hypothetical substance that doesn't emit, absorb, or reflect light, but interacts gravitationally. It was never detected directly; it was simply required to balance the equations.

A similar story unfolded decades later with cosmic expansion. Observations of distant supernovae in the late 1990s revealed that the universe's expansion wasn't slowing down, as gravity should cause — it was accelerating. Once again, instead of revisiting the underlying model of cosmic dynamics, an unseen component was added to account for the effect. This became dark energy, described vaguely as a repulsive force or vacuum energy driving space apart.

Both ideas were mathematical necessities first, physical realities second. Dark matter was the particle placeholder for missing gravitational effects; dark energy was the mysterious accelerator for unexplained expansion. And in both cases, the underlying assumption remained fixed: the gravitational model was correct — the universe must be hiding something to make it work.

Over time, these placeholders grew into dominant parts of the cosmological model. Today, textbooks state that roughly 27% of the universe is dark matter and 68% is dark energy, leaving less than 5% for all the ordinary matter we can actually observe. The irony is that these vast, unseen components were not found by looking outward — they were invented by looking inward, at equations that wouldn't balance without them.

6.2 The Mystery Particle Problem

Once the concept of dark matter was introduced to patch the gap in galactic rotation models, the next step seemed obvious: if it exerts gravity, it must be made of particles. This assumption launched a decades-long experimental hunt for the invisible mass.

The leading candidates were WIMPs — *Weakly Interacting Massive Particles*. These hypothetical particles were thought to be heavy enough to generate the required gravitational pull but interact so weakly with normal matter that they would be nearly impossible to detect. Physicists also proposed axions, ultra-light particles that could form diffuse clouds around galaxies. Other possibilities included sterile neutrinos, supersymmetric particles, and even exotic primordial black holes.

Billions of dollars have been spent on detection experiments:

- Deep underground detectors, shielded from cosmic rays, to catch rare particle interactions.
- Cryogenic arrays to sense tiny energy transfers from invisible collisions.
- Particle accelerators like the Large Hadron Collider to try to produce dark matter in high-energy collisions.

The result? No confirmed detection. Not a single candidate particle has been conclusively observed in nature.

In a healthy scientific process, this repeated null result might have prompted a re-examination of the original assumption: perhaps the missing mass isn't a particle at all, or perhaps gravity behaves differently on these scales. Instead, the absence of evidence has simply expanded the search. Every time an experiment fails to find the predicted particle, new variations are proposed: lighter WIMPs, heavier WIMPs, self-interacting dark matter, fuzzy dark matter — the list keeps growing.

This is where the logic turns circular:

- The particles must exist because the equations require them.
- If the experiments don't find them, the particles must be harder to detect than we thought.
- If they are harder to detect, that explains why we haven't found them yet — so the search must continue.

The possibility that the equations themselves might be incomplete — that the gravitational model or the interpretation of the data might be flawed — is rarely given equal weight. Dark matter has become so embedded in the standard cosmological model that its absence would be more disruptive to mainstream theory than its presence is to physical reality.

From a Universal Energy Dynamics (UED) perspective, this is a textbook case of chasing a mathematical placeholder as if it were a physical certainty. The "mystery particle" may never appear because the missing piece is not an invisible substance at all, but the ripple-tension behavior of the structured vacuum energy (SVE) — a phenomenon that changes how gravity manifests without needing new matter to explain it.

6.3 UED's Ripple-Tension Explanation

In the Universal Energy Dynamics (UED) framework, the anomalies in galactic motion and cosmic expansion are not signs of missing matter or mysterious energy — they are signs of field behavior we have been misreading.

At the heart of UED is the structured vacuum energy (SVE) — a universal field with inherent *ripple tension*. This tension is not static; it varies with local and large-scale alignment patterns. Matter, light, and energy are not moving through an empty void, but interacting constantly with this ripple-structured medium.

1. Galactic Rotation Without Hidden Mass

Mainstream models expect stars on the outer edges of a galaxy to orbit more slowly than those near the center, based on visible mass and Newtonian gravity. Observations show they orbit much faster than predicted. In UED terms, this is not because unseen mass is pulling them, but because ripple tension gradients in the SVE change the effective gravitational behavior.

- As the galactic ripple field aligns on large scales, tension becomes more uniform across the disk.
- This uniform tension reduces the expected decay of orbital velocity with distance from the center.
- The result matches the "flat" rotation curves — without adding an invisible halo of particles.

2. Cosmic "Acceleration" Without Dark Energy

Mainstream cosmology interprets the apparent acceleration of the universe's expansion as evidence for a mysterious, repulsive "dark energy" that counteracts gravity. UED offers a different interpretation: large-scale ripple field alignment periodically changes the apparent relationship between distance and velocity of distant galaxies.

- When ripple alignment increases over vast regions, signal travel times and observed separations can shift in ways that mimic acceleration.
- The effect is an *emergent property* of the field's geometry and tension — no extra cosmic ingredient is required.

3. Emergent Properties, Not New Substances

Both effects — flat rotation curves and apparent cosmic acceleration — emerge naturally from the same ripple-field principles:

- Ripple tension governs how energy-mass distributions interact at every scale.
- Alignment patterns determine how these interactions appear to observers.
- No hidden particles, no extra dimensions, no exotic forces — only a better understanding of the medium in which all interactions occur.

In this light, the dark matter/dark energy paradigm is revealed as a patchwork solution — adding imaginary components to balance equations instead of refining the field model. UED replaces these ad hoc additions with a unified physical mechanism that is measurable, falsifiable, and free from ghost entities.

6.4 The Fix-by-Addition Myth Factory

In ancient times, when a core belief faced a contradiction, the solution was rarely to replace the belief itself. Instead, the belief was protected — and something new was added to patch the gap. The heavens didn't move the way the theory predicted? Add another layer of invisible mechanism. The gods' behavior didn't match the observed world? Invent a new god, or a new myth, to explain why.

Modern cosmology, for all its mathematical sophistication, often follows the same pattern.

1. The Core Belief That Must Survive

In today's case, the core belief is the mainstream gravity model — general relativity as currently framed. This model is treated as too foundational to alter. When anomalies arise — such as unexpected galactic rotation speeds or an accelerating cosmic expansion — the instinct is not to reconsider the model, but to protect it at all costs.

2. Adding the Invisible

The fix comes in the form of unseen components:

- *Dark matter* to save gravitational predictions at galactic and cluster scales.
- *Dark energy* to preserve the expansion model without changing the underlying equations.

These entities are introduced not because they were discovered, but because they are needed to make the math work.

3. From Hypothesis to "Fact"

Over time, these placeholder concepts acquire an aura of reality. They get names, properties, and elaborate sub-theories. Entire research programs, experiments, and careers form around detecting them — despite no direct evidence. Much like ancient mythic additions, they transition from *temporary explanations* to permanent fixtures in the belief system.

4. Historical Parallels

- **Ether** in classical physics — a hypothetical medium for light waves, "proven" by its mathematical necessity, until experiments like Michelson–Morley dismantled it.
- **Crystal spheres** in ancient astronomy — added to preserve Earth-centered cosmology when planetary motion didn't match predictions.
- **Dark matter and dark energy** today — vast, invisible components added to balance the cosmic ledger without altering the gravitational model itself.

5. The Myth-Making Machine in Action

The danger of fix-by-addition is that it builds an increasingly elaborate structure on a possibly flawed foundation. The more layers added, the harder it becomes to question the base assumption. What begins as a patch becomes a worldview — a self-consistent but possibly ungrounded narrative, complete with its own high priests, sacred constants, and unobservable realms.

In UED terms, this is myth-making with equations instead of deities. The elegance of the math replaces the poetry of the verse, but the underlying process — preserve the dogma, add the invisible — remains unchanged.

6.5 Toward a Ripple-Field Cosmology

The promise of Universal Energy Dynamics is not just to explain *away* dark matter and dark energy, but to replace them with a single, physically grounded framework that unifies cosmic behavior from the scale of planets to the scale of the observable universe.

In ripple-field cosmology, all gravitational, rotational, and expansion-like phenomena emerge from the structured vacuum energy (SVE) and its ripple-tension dynamics — without resorting to unseen "stuff" that exists only because the math demands it.

1. Galactic Dynamics Without Hidden Mass

In this model, the unexpected flatness of galactic rotation curves is not a sign of invisible halos of exotic particles. Instead, it reflects variations in ripple-field tension across the galaxy's span. These variations alter the effective gravitational interaction without requiring extra matter — the effect is field-based, not mass-based.

2. Cosmic Expansion Without a Mystery Force

Similarly, the observation that distant galaxies seem to be accelerating away is reinterpreted as a large-scale field alignment phenomenon. Regions of the universe with different ripple-field configurations create the appearance of acceleration when observed from our local alignment frame. There is no need to postulate an anti-gravity force or "dark energy" fluid pervading all of space.

3. One Mechanism, Many Scales

This approach uses the same physical principle — ripple-field tension and alignment — to explain:

- Planetary orbits and stability.
- Spiral galaxy dynamics.
- Cluster interactions.
- Cosmic-scale motion patterns.

The same law applies, with no separate invisible substances for different scales.

4. Observational Tests to Break the Tie

Ripple-field cosmology makes distinct predictions that can be tested against dark matter/dark energy models:

- Rotation curve gradients should correlate with measurable variations in local field alignment, not with a universal "missing mass" ratio.
- Gravitational lensing patterns should reflect ripple tension gradients, and not require a uniform dark matter halo distribution.
- Cosmic microwave background (CMB) anomalies should align with predicted ripple-field regions, rather than being statistical noise.
- Local acceleration deviations in spacecraft trajectories or pulsar timing arrays could directly reveal ripple-

field asymmetries — effects that dark energy cannot predict.

5. From Patchwork to Principle

By moving away from additive fixes, UED re-centers cosmology on cause-effect grounded in measurable field dynamics. The structured vacuum becomes the active participant in shaping cosmic behavior, and alignment physics replaces invisible placeholders with testable, falsifiable mechanics.

Ripple-field cosmology doesn't just tidy up the cosmic ledger — it changes the bookkeeping system entirely.

Chapter 7 – Quantum Collapse

The Mystical Vanishing Act

You've heard this story before, though the costumes change. In one version, a crowd gathers in a temple, gasping as the stone idol appears to sip milk offered in devotion. In another, a group of physicists in a lab watches their detector light up, announcing that a particle — until just now "everywhere" — has suddenly "chosen" a single position the moment they looked.

Different settings, different languages, same dramatic reveal: an invisible process, hidden from direct scrutiny, is framed as a sudden, almost magical transformation. The audience is left with the same sense of awe — and the same unanswered question.

Both rely on a leap across an unlit gap. In religion, that leap is called faith. In physics, it's called "wavefunction collapse." In both, the curtain drops before we can see what's actually happening behind the scenes.

In UED, we don't stop at the curtain. We step through it.

7.1 The Orthodox Quantum Story

In the standard textbook version of quantum mechanics, the universe at the smallest scales is not made of definite objects in definite places, but of a strange, fluid mathematics called the wavefunction. Introduced by Erwin Schrödinger, this wavefunction isn't a physical wave rippling through space like sound or water — it's a mathematical construct that encodes the probabilities of finding a system in various possible states. It is, according to the orthodox view, the complete description of reality at that level.

Here's where the drama begins: as long as no one "looks," the wavefunction evolves smoothly and predictably, like a perfectly choreographed dance. But the moment a measurement is made, something abrupt and mysterious happens — the collapse. The wavefunction, once spread out over many possibilities, allegedly "chooses" a single outcome. A particle that was "everywhere" now exists "right there."

In the Copenhagen interpretation, the most widely taught version of this story, the observer plays a crucial role in this transformation. The act of measurement isn't just passive recording — it's said to somehow *create* the reality observed. Until then, the particle's properties are not merely unknown; they are undefined.

This narrative has been reinforced with vivid and often misleading imagery in popular science: electrons "popping" into existence when watched, photons "deciding" which path to take in a double-slit experiment, or the famous cat that is both alive and dead until someone lifts the lid. These images make quantum mechanics sound like a universe run on whimsy

— a place where nature waits for us to look before making up its mind.

It's an engaging story. But like most engaging stories, it leaves one crucial question unanswered: is this a description of what's really happening, or a myth dressed in mathematical robes?

7.2 The Miracle Problem

If you strip away the quantum jargon, "wavefunction collapse" has an unsettling resemblance to the classic storyteller's shortcut: *"And then a miracle happened."* In both cases, the narrative reaches a point where the mechanism is unclear, so an event is declared rather than explained. In religion, it's the divine hand intervening without traceable cause. In orthodox quantum mechanics, it's the "collapse" — an abrupt transformation of a probability cloud into a single, definite outcome — with no physical explanation for *how* or *why* it occurs.

The issue is not that the mathematical formalism is flawed; in fact, the Schrödinger equation is beautifully consistent in describing the smooth evolution of the wavefunction. The problem is that this formalism stops at the moment of observation. The collapse is *not* derived from the equation; it's bolted on as an extra rule — a postulate. There's a canyon-wide gap between the math (which says probabilities evolve) and the physics (which demands a real, definite outcome).

Different interpretations of quantum mechanics attempt to patch this gap, but each sidesteps rather than solves it:

- **Copenhagen Interpretation**: Avoids the "how" entirely, asserting that collapse is a fundamental process that occurs during measurement — and measurement just *is* what it is. No further questions, please.
- **Many-Worlds Interpretation**: Dispenses with collapse entirely by claiming all possible outcomes happen, each in its own branching universe. This replaces one miracle with an infinite number of unseen universes, each born without observable evidence.
- **Objective Collapse Theories**: Propose that collapse is a real, spontaneous event triggered by some physical threshold (mass, complexity, time), but these models remain speculative and unverified.

In every case, the core question — *What physically happens to turn a possibility into a reality?* — is never answered in a way that ties directly to observable cause-effect. Instead, collapse remains an "event" without a mechanism, a metaphysical sleight of hand that passes in physics only because it's hidden inside mathematical formalism.

7.3 UED's Reinterpretation: ATT Resolution

In the Universal Energy Dynamics (UED) framework, there is no mysterious, metaphysical *"collapse"* where a particle makes a last-second choice about its fate. What mainstream quantum mechanics treats as a sudden vanishing act is, in UED, a straightforward matter of alignment between fields — specifically, the observer's field configuration and the event's ripple pattern in the Structured Vacuum Energy (SVE).

When you measure something, you're not "forcing nature to decide." You're engaging in a process where your local field — along with any instruments you're using — must physically align with the ripple configuration of the system you're probing. This alignment is not instantaneous; it has a measurable, definable interval known as Alignment Threshold Time (ATT).

ATT is the minimum time required for that alignment to stabilize. Before ATT is crossed, what you have is incomplete coherence: the ripple information you receive is partial, fluctuating, and still embedded in multiple possible outcomes. Once ATT is reached, the alignment locks in, and you get a definite, reproducible outcome — not because the particle "chose," but because the configuration has fully resolved in your observational frame.

In this view:

- **Before ATT**: The event exists in ripple form, but your interaction with it is under the stability threshold. The outcome appears indeterminate because the ripple configuration hasn't stabilized relative to your field.
- **At ATT**: Field alignment reaches the threshold; the interaction becomes fully real in your reference frame. The "collapse" is simply the completion of alignment, not a metaphysical leap.
- **After ATT**: The measured state is fixed for you, but other observers in different field conditions may still

be in their pre-ATT phase — explaining why "when" a result is seen can vary without invoking many worlds or magical branching.

The elegance of this model is that the particle never makes a decision — it always exists as a ripple configuration with well-defined causal history in the SVE. The apparent discontinuity in mainstream quantum mechanics is nothing more than the observer hitting the alignment threshold that allows that configuration to be *experienced as definite*.

In short, UED turns "wavefunction collapse" from a mystical event into a causal, measurable, ripple-field process — no miracles required.

7.4 No Mysticism, Just Physics

One of the most persistent problems in quantum mechanics is its **anthropocentric bias** — the way popular and even formal interpretations center human consciousness as the key player. In the Copenhagen model, for instance, *"measurement"* often gets conflated with *"a conscious observer looking"*. This inadvertently smuggles a philosophical ghost into physics, turning a well-defined mathematical framework into a quasi-spiritual narrative.

UED strips this away entirely. In the ripple-field model, the universe doesn't care whether you're a PhD physicist, a lab robot, a bird flying through a magnetic anomaly, or a cosmic dust grain drifting through a gamma-ray burst — all are *observers* in the physical sense, because all have a field configuration that can align with ripple patterns in the Structured Vacuum Energy (SVE).

Here, **Alignment Threshold Time (ATT)** is the universal rule. It applies everywhere, for everything:

- A camera sensor hitting ATT records a definite photon position not because the universe recognizes "human intent," but because the sensor's field has aligned with the ripple configuration for long enough to stabilize the interaction.
- A speck of dust passing through an interstellar ripple pattern reaches ATT in the same way, even though it has no awareness at all — its atomic structure simply meets the physical alignment requirement.
- A living organism, whether human or insect, doesn't impose "special" quantum rules. Its measurements are just biochemical and electromagnetic field interactions obeying the same ATT principles.

Once ATT is reached, ripple causality takes over: the outcome distribution — the probabilities of where, when, and how events are realized — follows from the coherent chain of ripple interactions in the SVE. No "mystical choice," no "wave collapse triggered by human will," just structured energy dynamics producing consistent cause-effect sequences.

By replacing the "conscious observer" myth with a field-interaction requirement, UED not only removes the magical overtones from quantum theory but also gives us a measurable, testable pathway for studying what happens at the moment of apparent quantum decision. It's not the mind bending reality — it's the physics of alignment making reality visible.

7.5 Testable Differences

The power of UED's reinterpretation of quantum events lies not just in its philosophical clarity, but in its ability to generate predictions that measurably differ from standard quantum collapse models. In other words, this isn't just a thought exercise — it can be taken to the lab and challenged.

In the Copenhagen view, collapse is instantaneous and irreducible. You cannot, in principle, "peek" at the system between superposition and final outcome without disturbing it. In UED, however, there is a *transitional phase*: the pre-alignment state — the period between interaction initiation and the Alignment Threshold Time (ATT) being reached. This is not a metaphysical "limbo" of possibilities; it's a real, physical phase where ripple-field alignment is still building toward stability.

Predictions where UED differs from standard collapse models:

Outcome stabilization is time-dependent, not instantaneous.

- UED predicts measurable variations in outcome certainty if measurements are attempted *just before* ATT versus *just after*.
- Collapse models say no such gradation exists — you either have superposition or a final outcome.

Multi-observer effects depend on relative ATT windows.

- Two observers (or detectors) measuring "simultaneously" but with different field configurations could record outcomes at slightly different effective times.
- Collapse models expect identical results at the same spacetime coordinates.

Ripple interference can alter ATT without changing the underlying probabilities.

- Introducing controlled ripple distortions into the environment could shift the ATT without altering the overall distribution, revealing a physical mechanism behind outcome timing.

Possible laboratory experiments:

- **Delayed-Choice Interference with ATT Tracking:** Modify a standard delayed-choice quantum eraser experiment to introduce tunable ripple distortions (e.g., electromagnetic or acoustic fields affecting alignment speed). Track whether interference patterns vanish or persist at different ATT intervals.
- **Multi-Observer Timing Studies:** Place two detectors with different physical or electromagnetic configurations in the same measurement path. Observe whether their outcome recording times drift

relative to one another in a way correlated to ATT predictions.

- **Pre-Alignment Probing:** Develop ultrafast probing techniques to "sample" the system before ATT is reached, checking for incomplete stabilization signals — a signature of the pre-alignment phase that collapse theory says should not exist.

What this would reveal: If experiments detect systematic, measurable variations in the time to outcome stabilization, or observer-specific differences in outcome timing, it would point strongly toward the ATT model and away from metaphysical collapse. In short, it would demonstrate that the so-called "mystery of quantum measurement" isn't about the universe waiting for someone to look — it's about the physics of ripple-field alignment making the event real.

7.6 From Quantum Paradox to Ripple Logic

By stripping away the mystical aura surrounding "wavefunction collapse," UED transforms many of quantum mechanics' most celebrated paradoxes from unsolvable riddles into straightforward field-alignment scenarios.

Take **Schrödinger's cat**. In the orthodox view, the cat is suspended between life and death until someone opens the box, as if reality itself were holding its breath. Under UED, there is no such suspended limbo. The cat's condition — alive or dead — is determined once all relevant ripple interactions within the box reach their **Alignment Threshold Time**

(ATT). The outside observer's measurement merely aligns *their own* ripple field with that pre-existing, fully resolved configuration. The drama is not in the physics but in the observer's delayed awareness.

Similarly, **Wigner's friend** — the idea that different observers could see different "versions" of reality — collapses under ripple logic. In UED, differing observations arise only if observers' fields have different ATT windows. Once the full system aligns, every observer interacting with it thereafter will encounter the same configuration. There is no permanent split of realities — just staggered moments of alignment that give the illusion of conflicting truths.

This also reframes quantum uncertainty. The Heisenberg principle, often portrayed as proof of nature's inherent randomness, becomes in UED a statement about incomplete alignment: before ATT, the ripple field does not yet hold a fully resolved configuration for all measurable parameters. Attempting to measure too early forces a partial interaction that obscures complementary variables — not because nature is unknowable, but because the alignment is physically incomplete.

Most importantly, this ripple-based reinterpretation builds a seamless bridge between the quantum and the macroscopic. The same logic that explains an electron's "collapse" explains why a car engine won't start until all its mechanical, electrical, and environmental ripples are in alignment. There is no mysterious cutoff point where physics switches from "weird quantum rules" to "normal classical cause-effect." It's all ripple logic — from subatomic particles to planetary systems —

governed by the same universal law of alignment and consequence.

In essence, UED does not just solve the measurement problem — it reintegrates reality. It turns paradoxes into process, and mystery into mechanics, without shrinking the wonder of physics.

Chapter 8 – Inflation and the Instantaneous Universe

Imagine the opening scene of the greatest blockbuster ever made. The lights go out, the music swells, and in less than a blink, the screen explodes into a panorama of galaxies, stars, and space-time itself unfurling in every direction. The universe — in full Hollywood style — has just made its grand entrance, roaring from nothing to everything faster than light could even keep up.

That's the cosmic inflation narrative we've been sold: a universe so impatient it hit the gas before the laws of physics could strap in. But was the early universe really in that much of a hurry... or have we simply mistaken *alignment* for *motion*?

In this chapter, we slow the reel, strip away the special effects, and ask a quieter, more radical question: what if the "instantaneous" universe didn't race anywhere at all — it just *came into focus*?

8.1 The Mainstream Inflation Story

In the standard cosmological script, the Big Bang wasn't the whole story — it was just the opening act. The real spectacle, we are told, came a fraction of a fraction of a second later, in a moment called *inflation*.

This idea didn't emerge because scientists were bored; it was built to patch three awkward holes in the original Big Bang model:

- **The Horizon Problem** – Why does the universe look so uniform in temperature, even in regions that should never have been in contact according to light-speed limits?
- **The Flatness Problem** – Why does the geometry of the universe appear so precisely balanced, as if it were fine-tuned on a cosmic knife-edge?
- **The Monopole Problem** – If the early universe was hot enough to create exotic magnetic monopoles, where are they now?

Inflation theory attempts to solve all three by proposing that, in the first sliver of a second after the Big Bang, the universe went through an ultra-fast expansion phase — stretching space itself at a rate far beyond the speed of light. And because, according to relativity, it's not *objects* moving through space but space itself expanding, this apparent faster-than-light growth supposedly doesn't break the cosmic speed limit.

In popular illustrations, it's shown as a balloon being blown up at impossible speed — galaxies and energy fields racing away from each other in a blur, leaving an image of instant grandeur and effortless magic. The sheer audacity of the concept makes it irresistibly cinematic: the cosmos as a stunt performer pulling off the ultimate opening trick.

Underneath the spectacle, however, is a quiet assumption — that the solution to early-universe puzzles is more speed, more scale, more drama. But as UED will later argue, the real solution might have nothing to do with any race at all. It may be that nothing "rushed" anywhere — the alignment simply settled everywhere at once.

8.2 The Core Assumptions Behind Inflation

Inflation wasn't born out of direct observation — it was reverse-engineered from mathematical necessity. Once the cosmic microwave background (CMB) was mapped in exquisite detail and shown to be astonishingly uniform in temperature, standard Big Bang models struggled to explain it. In those models, light and heat could not have traveled far enough in the available time to "even out" the early cosmos. Yet the CMB suggested an uncanny smoothness, as though the entire universe had been stirred in the same pot before cooling.

The solution proposed by Alan Guth and others was inflation — an extremely brief but stupendously fast period of expansion that would, in effect, erase initial irregularities and stretch the early universe into its current form. To make this work on paper, cosmologists introduced new conceptual players: the "inflaton" field and its associated particle. This hypothetical field would have dominated the universe's energy content for a fleeting instant, acting like a runaway accelerator until, for reasons still debated, it "switched off" and allowed normal expansion to resume.

But here's the catch:

- These fields and particles have never been observed in nature or in any experiment.
- They exist entirely as theoretical constructs to make the math balance — much like dark matter and dark energy in later cosmology.
- The parameters of the inflaton can be tuned to produce almost any desired outcome, making the theory remarkably flexible and frustratingly unfalsifiable.

Adding to the problem is the self-sealing nature of the inflation hypothesis. If inflation occurred, it would have wiped away any trace of what came before it, smoothing out not only matter but also information. That means there's no independent, observational access to pre-inflation conditions. In practice, we can't "look behind the curtain" — we can only model it, and the models are shaped to fit what we already think must have happened.

It's a bit like reading the first chapter of a book, finding an inconsistency, and then deciding there must have been a secret prologue — but one that no one can ever read, and whose contents can be rewritten at will to explain whatever you like.

8.3 UED's Alternative: Coherence, Not Hyperdrive

From the UED perspective, the early universe didn't need a faster-than-light "hyperdrive" to achieve its remarkable uniformity. The mistake lies in imagining the universe as a mechanical balloon that had to stretch its surface at impossible speeds to bring distant points into harmony. Instead, UED treats the early universe as a ripple-field coherence event — a phase in which alignment, not motion, set the initial conditions.

In this model, time is measured as Alignment Threshold Time (ATT), the interval required for a ripple configuration to stabilize and become causally effective. In the moments after the initial ripple of creation, the Structured Vacuum Energy (SVE) — the underlying medium of all ripple patterns — allowed the alignment of distant regions before those regions became fully "separate" in a causal sense. This wasn't a case of matter racing through space; it was the entire field settling into a coherent configuration at once, like a vibrating drumhead finding a single resonant mode.

The uniformity of the cosmic microwave background (CMB) is no mystery here. It doesn't require space to have physically expanded faster than light, because no exchange of photons across vast distances was necessary to establish the observed symmetry. The "sameness" comes from simultaneous field alignment — the ripple patterns everywhere in the early SVE matched up before ATT was crossed for those patterns to lock into place.

From this vantage point:

- Distant regions appear uniform not because they traded information at impossible speeds, but because they were never truly out of sync to begin with — they emerged from the same coherence event.
- There is no need for hypothetical inflaton fields or exotic "kick-start" mechanisms. The observed balance is an intrinsic property of how ripple fields stabilize within SVE.
- The inflation paradox dissolves — the early cosmos didn't "rush" anywhere; it simply aligned everywhere.

In short, UED replaces the drama of a cosmic drag race with the elegance of a perfectly tuned orchestra: each note struck everywhere at once, resonating into the universe we now observe.

8.4 Dissolving the Horizon Problem Without Inflation

The so-called *horizon problem* is framed in mainstream cosmology as a communication crisis: how could two regions of the early universe, now separated by billions of light-years, have "talked" to each other fast enough to agree on the same temperature and density? Within the standard model, there's only one fix — assume the universe once inflated faster than the speed of light, letting distant points once be in close contact before rushing apart.

UED reframes this entirely. The real question is not *how did they communicate fast enough?* but *how did they align?* Communication assumes a message traveling from A to B at some speed. Alignment is a different game altogether — it is a field property, not a signal transaction.

Ripple fields embedded in the **Structured Vacuum Energy (SVE)** don't require a photon or particle to cross space for coherence to emerge. Coherence is an immediate state — if the underlying field enters a stable ripple pattern, every location within that field shares that pattern instantly. This is not magic or nonlocal "spooky action"; it is simply the way structured fields behave when they cross an **Alignment Threshold Time (ATT)**.

ATT acts as the stabilizing gate:

- Before ATT is reached, ripple patterns may be in flux, misaligned, or in competing modes.
- When ATT is crossed, the configuration locks in — the same way a vibrating plate abruptly "decides" on a single standing wave pattern across its surface.
- This lock-in can happen over the entire accessible field simultaneously, regardless of spatial separation, because no *signal* is being sent; the field itself is resolving into a coherent state.

In this view, the CMB's uniformity is no longer a puzzle about faster-than-light travel. The entire early universe was a single, still-cohering ripple field, and its uniformity is simply the imprint of that coherence event. No inflationary burst, no hypothetical inflaton particle — just a natural property of SVE ripple alignment.

What cosmologists mistook for a need for "super-speed" was really a case of misreading alignment as motion. The horizon problem dissolves the moment we stop forcing light-speed limits onto processes that never involved light travel in the first place.

8.5 Observational Reinterpretations

If UED's alignment-based model is correct, much of what cosmology now attributes to "early universe inflation" can be reinterpreted as the natural outcome of ripple-field phase coherence in the Structured Vacuum Energy (SVE).

CMB uniformity

Under standard inflationary theory, the near-perfect uniformity of the cosmic microwave background (CMB) is explained by a brief period when the universe's expansion was exponentially fast, allowing once-connected regions to retain the same temperature even after being driven far apart.

UED flips this logic: the uniformity isn't a relic of spatial proximity — it's the signature of a field-phase lock-in that occurred before causal separation was even meaningful. The CMB's smoothness is the "frozen" ripple pattern from when the SVE field crossed its Alignment Threshold Time (ATT), imprinting a coherent state across the entire observable domain in one instant.

Large-scale structure formation

Mainstream models attribute the seeds of cosmic structure to quantum fluctuations in the inflaton field, stretched across the universe during inflation. In UED, the density variations that later grew into galaxies and clusters arise instead from ripple-tension gradients in the SVE. These gradients are not random noise amplified by inflation, but emergent differences in ripple compression and alignment stability — a more physical and measurable origin for structure.

Baryon Acoustic Oscillations (BAOs)

BAOs are often described as the "sound waves" from the early plasma, their scale preserved by the inflationary expansion. UED reinterprets them as standing ripple modes in the early SVE-matter coupling, where the characteristic scale is set not by how far waves could "travel" before recombination, but by the natural geometry of the ripple field under its maximum coherence. This reframing shifts the focus from "sound travel distance" to field geometry and tension nodes, offering new ways to model and test the BAO signal.

The shift in interpretation

By removing the requirement for a superluminal expansion phase, UED simplifies the story of the early universe. The same observations — CMB uniformity, structure seeds, BAOs — are not remnants of a fleeting hyperdrive moment, but the inevitable consequences of field alignment physics.

This also means that future measurements don't need to hunt for an inflaton particle or inflation's "smoking gun." Instead, they can test ATT lock-in signatures, ripple-tension patterns, and coherence decay rates — all of which are physically measurable without invoking hypothetical substances or forces.

8.6 Testable Differences Between Inflation and UED Coherence

One of the most persistent criticisms of inflation theory is that it is notoriously hard to falsify — almost any observed feature in the early universe can be accommodated by tweaking the inflationary potential, the duration of the expansion phase, or the nature of the hypothetical inflaton field.

UED coherence, in contrast, makes specific, non-adjustable predictions that can be confronted with data and even explored in controlled physical analogues.

CMB anisotropy patterns without inflation

If the CMB's uniformity is the result of field-phase alignment rather than superluminal stretching, then the fine-scale anisotropy patterns should bear geometric and harmonic relationships consistent with ripple-mode coupling, not quantum fluctuations stretched beyond the horizon. In practice, this means:

- The angular power spectrum would exhibit subtle phase-locked mode correlations absent in inflationary randomization.
- The statistical distribution of anisotropies would deviate from pure Gaussianity in ways predicted by ripple-tension coupling equations.
- Certain correlations should persist across multiple angular scales, reflecting the global coherence at ATT lock-in rather than stochastic noise.

Laboratory analogues

While we cannot rewind the universe, UED's ripple-phase alignment physics is not unique to cosmology. Potential experimental platforms include:

- Bose–Einstein condensates where collective

wavefunction alignment mimics field coherence without particle exchange.

- Shallow fluid surface wave systems that can be driven into large-scale phase coherence without directional propagation.
- Superconducting arrays where phase-locking occurs across the lattice almost instantaneously due to underlying field coupling.

These analogues could provide measurable insights into how coherence emerges and decays, offering scaling laws that can be compared to cosmic data.

Astronomical searches for coherence decay

In UED, the post-ATT universe does not remain perfectly aligned forever — ripple-tension gradients and large-scale structure growth gradually erode the pristine coherence of the early field. This predicts that:

- The oldest light (CMB) should show slightly higher coherence than younger cosmic signals such as 21-cm emission from the cosmic dawn.
- Coherence decay rates can be compared across different cosmic epochs to test the ripple-field dynamics directly.
- No inflation "relics" (e.g., primordial gravitational waves of inflationary origin) should be found — but coherence-geometry relics may exist and could be detected with next-generation sky surveys.

The falsifiability advantage

Where inflation can always invent a new inflaton potential to match the data, UED coherence commits to concrete observational signatures and measurable decay behaviors. This transforms the origin of cosmic uniformity from a speculative hyperdrive story into a testable physics problem — one that can be probed from the lab bench to the cosmic horizon.

8.7 Why the Hyperdrive Narrative Persisted

The idea of cosmic inflation has endured not only because of its mathematical elegance within the standard cosmological model, but also because it satisfies deeper psychological and institutional desires in science.

The lure of a cinematic beginning

Human storytelling instincts favor dramatic origin tales — moments of sudden transformation, powerful enough to imprint themselves on the imagination. The inflation narrative is essentially a cosmic action sequence: in less than a trillionth of a second, the universe grows from subatomic to astronomical scales, faster than light itself.

This imagery is intoxicating. It evokes grandeur, speed, and inevitability. For scientists and the public alike, it's far more gripping than a slow, invisible process of ripple-field coherence. Inflation lets cosmology start with a bang-plus-a-hyperdrive, a spectacle that feels worthy of the "birth of everything."

Protecting the core dogma

From a structural perspective, inflation became the reinforcing scaffold for the Big Bang framework. When observations like the CMB's uniformity or the large-scale distribution of galaxies didn't fit the original model, inflation provided a way to "explain away" these mismatches without discarding the foundation.

This aligns with a historical pattern in science: when a beloved theory is challenged, the first instinct is to append new components rather than rebuild from first principles. Inflation thus functioned as a patch, elegantly formulated but still fundamentally designed to keep the original edifice standing.

The untestable elegance problem

While framed as a scientific theory, inflation's lack of independent pre-inflation observables makes it extraordinarily difficult to falsify. Almost any feature in the CMB can be reinterpreted by adjusting the shape of the inflaton potential, the duration of the inflationary phase, or the details of reheating. This flexibility is a double-edged sword: it makes the model adaptable, but also immunizes it from failure — a hallmark of what UED identifies as myth-preserving physics.

The special effects budget of cosmology

In blockbuster films, when the script falters, a lavish special effects sequence can distract the audience and carry the narrative forward. In modern cosmology, inflation often plays the same role. It fills the gap between the unobservable Planck era and the first measurable light with a burst of mathematically spectacular, visually irresistible speculation.

This spectacle is compelling not because it's directly evidenced, but because it looks impressive in simulation and feels intuitively satisfying — a storytelling flourish dressed as inevitability.

8.8 Closing Reflection – Instantaneous Without Rushing

The picture painted by UED is radically different from the mainstream's cosmic hyperdrive. The early universe did not sprint into existence, burning through some unimaginable burst of momentum. It did not tear through space at impossible speeds or balloon in a desperate race to spread itself thin.

Instead, it aligned.

In the UED framework, the "moment" of cosmic birth was not an explosion of distance, but an immediate ripple-field coherence event within the Structured Vacuum Energy (SVE). Every nascent region of what we now call the universe was already part of the same underlying field. There was no "communication problem" to solve because there was no separation yet in the ripple phase — everything was already connected through the deeper substrate.

This alignment was instantaneous not because anything moved faster than light, but because there was nothing to move. Coherence is a condition, not a journey. In this light, the universe's first breath was not a breakneck race through space, but the settling of an entire field into a stable initial configuration.

This reinterpretation is more than a technical correction. It reframes our sense of beginnings: the profound can happen without spectacle, without haste, without even motion in the conventional sense. What inflation describes with speed, UED describes with order.

And this is not unique to cosmology. The same principle — where apparent "instant" events dissolve into ripple logic — appears elsewhere in physics: quantum state resolution, field synchronization in plasma, even certain macroscopic phase transitions.

In the next chapter, we will apply UED's Reality Filter to these other "miracle moments" in science, stripping away the illusion of sudden magic and revealing the quiet, structured processes beneath.

Chapter 9 – Multiverses and Extra Dimensions

Math's Pantheon

In the ancient world, people looked up at the mountains and imagined gods living above the clouds — invisible realms where mortal rules did not apply. Today, many physicists look not to mountaintops but to equations, and from them conjure new deities: ten-dimensional strings, parallel universes, and entire cosmic neighborhoods that can never be visited. The scenery has changed from marble temples to chalkboards, but the reverence feels familiar.

The question is as sharp now as it was in antiquity: are we truly uncovering hidden layers of reality, or have we built another pantheon — this time not of gods, but of mathematical constructs — worshipped for their elegance rather than their evidence? In the age of precision instruments and cosmic telescopes, the most extravagant worlds still seem to live entirely on paper. And in that gap between mathematics and measurement lies the heart of our investigation.

9.1 The Rise of the Dimensional Gods

The modern fascination with extra dimensions began as a quiet mathematical curiosity. In the 1920s, Theodor Kaluza and Oskar Klein proposed that by adding a fifth dimension to Einstein's general relativity, one could elegantly merge gravity with electromagnetism. The extra dimension was invisible, they argued, because it was compact — curled up so tightly that no experiment could detect it. This was not science fiction yet; it was a clever trick to make equations cooperate.

Decades later, the idea reemerged in string theory, where the "strings" — one-dimensional vibrating filaments — required not just one hidden dimension, but six or seven beyond the familiar three of space and one of time. In these theories, the extra dimensions are not optional—they are mathematically mandatory for the equations to remain consistent. Like a finely tuned symphony, the mathematics sounds "right" only if these unseen dimensions are assumed to exist.

What began as niche theoretical work quickly seeped into the broader physics culture. The sheer elegance of the mathematics, combined with its promise of unifying all forces and particles into a single grand framework, was intoxicating. Even without experimental proof, higher dimensions became a badge of intellectual sophistication — a realm where the initiated could speak of Calabi–Yau manifolds and brane-worlds with the same reverence priests once reserved for divine mysteries.

By the time the concept reached popular science media, it had been stripped of its mathematical caution and dressed in cinematic grandeur: portals to parallel worlds, universes stacked like pages in a book, strings vibrating in hidden chambers of reality. These "dimensional gods" were no longer shy, compact mathematical curiosities — they had become central characters in the modern mythos of physics.

In essence, extra dimensions migrated from blackboards to bestseller lists, propelled not by direct measurement but by the allure of a perfect equation — an invisible Olympus built entirely out of mathematics.

9.2 Multiverse Mania

The modern "multiverse" is not one idea but a sprawling family of speculative frameworks, each born from a different corner of theoretical physics — and each offering its own flavor of infinite reality.

Level I – Infinite Space Beyond Our Observable Patch

In the simplest version, our universe is just a small region of a much larger, infinite cosmos. The same physical laws apply everywhere, but beyond the limits of our observable horizon lie endless other regions with different configurations of matter. In this view, the multiverse is not exotic — it's just an extension of ordinary space that we can't access because light hasn't had enough time to reach us. This stems directly from the cosmological principle and general relativity, but it's still an inference — no one has seen "beyond the horizon."

Level II – Bubble Universes from Eternal Inflation

Here, the inflationary model of the early universe becomes a cosmic foam machine. Inflation doesn't just stop after creating our universe — it keeps going in other regions, spawning countless "bubble universes," each potentially with its own constants, particle types, or even dimensional structures. This idea comes from extending inflation theory beyond its initial role as a fix for the Big Bang's early puzzles, turning a convenient expansion phase into a perpetual generator of new worlds.

Level III – Many-Worlds Quantum Branching

Born from Hugh Everett's alternative to wavefunction collapse, the Many-Worlds Interpretation says every quantum event spawns a branching of reality — each possible outcome is realized in its own universe. In this vision, there isn't a single "you," but countless versions living out different histories. The appeal here is philosophical as much as physical: it removes randomness from quantum mechanics, but at the cost of multiplying realities into infinity.

Level IV – Mathematical Structures as Universes

This is the most abstract — proposed most famously by Max Tegmark. If mathematics is the ultimate foundation of reality, then every mathematically consistent structure "exists" as a universe in its own right. Our universe is just one such structure, inhabiting the grand "mathematical multiverse." This is not physics so much as a metaphysical declaration, where existence equals mathematical possibility.

Across all these versions, one fact stands out: none of them arise from direct observation. Each is the logical or philosophical overflow of an existing theory — an attempt to make the math or the interpretation "complete." They are intellectual extensions, not empirical necessities, often sustained more by elegance and narrative appeal than by data.

9.3 The Evidence Problem

Despite the bold confidence with which multiverses and higher dimensions are discussed in popular science outlets, the hard truth is simple: there is no direct detection of either. Not a single experiment has yielded a definitive signal that extra spatial dimensions exist, nor has any observation revealed the presence of a parallel universe brushing against our own.

Instead, the case is built on indirect inference — patterns, anomalies, or gaps in our current models are sometimes interpreted as "hints" of hidden realms. For example, slight irregularities in the cosmic microwave background, unexplained particle physics results, or mathematical neatness in unification models are occasionally spun into suggestive signs that we are touching the edge of a larger reality. Yet in each case, these interpretations rest heavily on pre-existing theoretical frameworks rather than emerging organically from the data itself.

The deeper issue is the self-sealing logic at the heart of many such claims. When no confirming signal appears, the absence of evidence is often reframed as *proof* of why these realities are beyond our reach. "We can't detect them because they are inherently undetectable" becomes the intellectual shield, turning unfalsifiable propositions into protected beliefs. The conversation then shifts from science to speculation — from hypotheses that invite testing to conceptual castles that float safely beyond the reach of empirical scrutiny.

This is not a minor quibble about evidence; it's a fundamental epistemic problem. A theory that can never be confirmed or refuted by observation or experiment may be interesting, elegant, and even internally consistent — but it no longer lives in the realm of physics. It becomes a story told in mathematics, with all the strengths and weaknesses that come with that.

9.4 UED's Physical Reality Filter

In Universal Energy Dynamics (UED), the threshold between *mathematical possibility* and *physical reality* is sharp and uncompromising: if no measurable ripple exists, it is not physically real. This is not a philosophical preference — it is a direct consequence of how UED defines the universe.

Every physical phenomenon, from the smallest particle interaction to the largest galactic movement, is ultimately expressed as a disturbance or alignment within the Structured Vacuum Energy (SVE). This "ripple signature" is not a metaphor — it is a measurable effect in the field. Even events we cannot observe directly, such as distant black hole mergers or particle interactions outside our immediate detection range, still leave a traceable imprint in the ripple field. These imprints are what make them part of the same physical reality we inhabit.

By this standard, extra dimensions that never interact with the SVE — producing no detectable phase shifts, tension gradients, or alignment effects — are simply mathematical abstractions. They may be elegant solutions in a theoretical equation, but without ripple evidence, they have no claim to existence in the physical domain. UED sees them as coordinates on paper, not landscapes in the universe.

The same filter applies to multiverse models. A "parallel universe" that never produces an interaction, no matter how subtle, with our ripple field is indistinguishable from nonexistence. If no causal or measurable link can be established, it fails the Physical Reality Filter.

This grounding in SVE-based physics is what keeps UED anchored. It demands that reality be tied to observable field effects — no matter how faint, delayed, or indirect — rather than to conceptual extrapolations that float free of empirical testing.

In other words, the Physical Reality Filter works like a cosmic customs checkpoint: if your phenomenon doesn't bring a ripple signature, it doesn't get to enter the universe.

9.5 Ripple Limits and the Illusion of Elsewhere

In UED, the Structured Vacuum Energy (SVE) is not a stage on which the universe plays out — it *is* the universe's substrate. Every star, atom, thought, and force is a ripple configuration within this single, continuous field. This has a profound implication:

There is no physical "outside" for an alternate universe to occupy. The ripple field is not contained in anything larger; it is already the whole framework of reality. Asking where another universe "is" becomes as meaningless as asking what's north of the North Pole.

When viewed through this lens, the idea of parallel realities collapses into two categories:

1. **Phase Variations in the Ripple Field** – Differences in ripple alignment can create *apparent* disconnections between regions, making them seem isolated or "parallel" to each other. However, these are not separate universes — they are parts of the same field temporarily misaligned, much like different wave patterns in the same ocean that seem unrelated until they interact.
2. **Incomplete Alignment Perception** – The human (or instrumental) observer may only be tuned to a subset of ripple configurations, leading to the impression of "hidden" worlds. These "elsewheres" exist only as *unrealized potential alignments* within the field, not as independent domains with their own physical SVE.

The illusion of elsewhere emerges when theory treats these effects as wholly separate realities rather than as field states within one continuous whole. Multiverse models that place universes in disconnected "bubbles" or "branes" inadvertently assume a container larger than the SVE — a concept that UED rejects as physically meaningless.

In short, the ripple field is self-contained and complete. There can be many configurations within it, but no *true* "other" universe standing apart from it. What mainstream cosmology calls parallel universes, UED reclassifies as phase-shifted domains of the same reality, no more separate than a ripple's crest is from its trough.

9.6 Why These Models Persist

Despite the absence of direct evidence, multiverse and extra-dimension models continue to thrive — not because of their empirical success, but because of their *narrative, cultural, and institutional momentum*.

First, there is the cultural and funding appeal. Grand speculative frameworks are marketable in both academic and popular circles. They promise the ultimate scientific prize — *explaining everything* — while offering tantalizing mysteries that can sustain decades of research proposals. A theory that hints at infinite worlds or hidden dimensions is easier to package for conferences, documentaries, and grant committees than one that quietly removes such extravagances. It taps into the same appetite that drives people to explore the deep ocean or climb the highest mountains: the desire to discover what lies beyond the visible frontier.

Then comes the psychological allure of "somewhere else". Human imagination has always been drawn to unseen worlds. From the underworlds and heavens of mythology to the enchanted forests of folklore, the idea that reality contains hidden realms is deeply embedded in our collective storytelling. Multiverse theories feel like the modern, physics-branded continuation of this tradition — swapping gods and demons for strings, branes, and bubble universes. This narrative resonance makes them feel "right" at an emotional level, even when the physics is speculative.

Finally, there is the prestige of mathematical beauty. Many of these models arise not from observation but from theoretical elegance — equations that unify forces or resolve inconsistencies *if* additional dimensions or universes are assumed. In physics, a theory that is mathematically graceful can earn admiration even if it has no clear path to experimental verification. This aesthetic bias means that a beautiful equation is often allowed to stand — and to be funded — long before nature has confirmed it.

The result is a self-sustaining ecosystem: prestige and promise outweigh empirical restraint. The absence of evidence doesn't shrink these models; instead, it's reframed as proof of their sophistication — *"Of course we can't detect them yet; they exist beyond our reach"*. This rhetorical shield ensures their survival, even as more grounded, testable frameworks like UED are sidelined for being less fantastical.

9.7 Testable Boundaries

If extra dimensions or parallel universes are to move from the realm of speculation into the domain of physics, they must leave a measurable footprint in the ripple field. UED draws an uncompromising line here: no ripple, no reality. This is not philosophical minimalism — it's an operational standard. Reality, in UED terms, is whatever produces *detectable and repeatable field effects within the Structured Vacuum Energy (SVE).*

From this standpoint, the only path forward is experimental. Instead of constructing ever more ornate mathematical worlds, the focus shifts to designing UED-based experiments that probe for measurable consequences of hypothetical extra dimensions. For example:

- **Ripple-Distortion Mapping**: If higher dimensions influenced our 3D ripple field, their coupling should subtly distort ripple tension patterns under specific configurations. Highly sensitive ripple-phase detectors could search for these deviations.
- **ATT Variation Tests**: Alignment Threshold Time could shift if information exchange is influenced by pathways beyond known space. UED experiments could test for these anomalies across extreme conditions — high-energy particle collisions, gravitational wells, or quantum interference setups.
- **Interference Persistence Studies**: Certain multiverse models imply leakage or cross-interaction between "branches." If such effects were real, they would appear as long-lived, low-level coherence anomalies in controlled ripple-interference tests.

The beauty of this approach is its clarity of outcome. If hypothetical extra dimensions exist in any physically meaningful way, they must interact with our measurable field structure. That interaction would leave consistent, reproducible ripple signatures — deviations in tension, alignment, or coherence that no 3D-only model could explain.

By contrast, if exhaustive, high-precision UED tests reveal no such ripple anomalies, then the verdict is simple: those dimensions may be beautiful in equations, but they are mathematical abstractions, not physical realities.

This makes ripple-detection thresholds the ultimate filter — a boundary that no speculative theory can cross without earning the right to be called physics. In UED, the test is not "Can you imagine it?" but "Can you measure it?"

9.8 Closing Reflection – The Temple and the Telescope

Mathematics can be a magnificent temple — an awe-inspiring structure built from axioms and logic, with archways of symmetry and domes of pure abstraction. Inside, the air is still and perfect, untouched by the noise of experiment. It is a sanctuary for the mind, a place where beauty and consistency can exist without compromise.

But in physics, a temple without windows is a prison. Equations, no matter how elegant, must eventually open their shutters to the sky and let reality's light pour in. Without that exchange, the most splendid mathematical architecture risks becoming a cathedral to itself — worshipping its own forms rather than the universe it claims to describe.

In UED, the only doorway into physical reality is through the ripple field's acknowledgement. A dimension or universe has no passport unless it can cross that border by producing a measurable ripple signature in the Structured Vacuum Energy. No imprint, no admission. It is a blunt, almost impolite standard — but it prevents physics from drifting into pure mythology with better equations.

The telescope, not the temple, decides what is real. And in UED's view, the final frontier of physics is not "elsewhere" — not in unreachable realms stacked like cosmic luxury condos beyond our perception — but right here. The secrets we chase in higher dimensions may already be encoded in the patterns and tensions of our own ripple field, waiting for us to read them.

The next chapter turns directly toward that challenge: the search for those unread patterns, the overlooked structures that have been hiding in plain sight, not in someone else's universe, but in our own.

Interlude – The Beauty and the Blindness

Part II has been a walk through physics' most celebrated theatres — each one staging a performance so spectacular that the audience forgets to check what's behind the curtain. The scripts are polished, the actors (mathematics, metaphors, and models) deliver their lines flawlessly, and the critics cheer. But beneath the applause lies a shared flaw: these stories survive not because they have been proven, but because they have been preserved.

From time travel loops to black hole infinities, from invisible cosmic "stuff" to quantum miracles, from universe-wide hyperdrives to infinite stacks of parallel worlds — the pattern repeats. A problem appears. The core model is too sacred to question. So, rather than rethinking the foundation, we decorate it with ever more elaborate scaffolding. And if the scaffolding sways, we add invisible beams. If it still creaks, we declare those beams unreachable by experiment.

UED refuses this theater's rules. It demands that the set be lit from all sides, that every prop be weighed, that every actor cast a measurable ripple in the Structured Vacuum Energy. No ripple, no reality — no matter how beloved the role or how mathematically handsome the face.

What this part has shown is not that mathematics is wrong — but that mathematics, unanchored, can become a myth machine as potent as any ancient cosmology. The next part will shift from dismantling these grand illusions to confronting the real stage: not the imagined "elsewheres" or unreachable "before-times," but the physical ripple field we are living in right now — and the patterns we have yet to read within it.

Part III – Beyond the Blackboard

Chapter 10 – The Seduction of Elegant Lies

Human perception is wired for patterns. From the swirl of a seashell to the symmetry of a snowflake, we instinctively find comfort in neatness. Symmetry, repetition, and closure give the brain a sense of completion — a signal that "this is right" simply because it fits together well. This tendency, however, is double-edged. It is as capable of leading us toward real discoveries as it is of luring us into elegant dead ends.

The trouble begins when elegance becomes a substitute for truth. A theory or model that feels aesthetically satisfying can easily be mistaken for reality itself. Its tidy equations, perfect symmetries, and closed loops deliver a psychological pleasure that makes it harder to question. We become more attached to how it *looks* than to whether it actually *works*. When reality refuses to conform to these beautiful structures, it is the data — not the design — that we tend to doubt.

This seduction isn't new. Ancient cultures framed the cosmos in perfect circles and harmonious ratios, not because nature demanded it, but because human cognition did. Modern science, despite its tools and methods, still falls prey to the same instinct. The danger is that we can build intellectual temples so compelling that they block our view of the outside world.

From the UED perspective, this is more than a philosophical caution — it is a warning about how theories can drift away from the field they claim to describe. In the previous chapter, we rejected unmeasured dimensions, not out of cynicism, but because the ripple field gave no sign of them. Yet this rejection raises an uncomfortable question: if these dimensions have no measurable presence, why were they embraced so passionately in the first place? The answer lies not in physics, but in psychology — in our attraction to the beauty of closed systems and the seduction of elegant lies.

10.1 The Ancient Precedent

Long before particle accelerators and space telescopes, humans built their understanding of the cosmos with a different kind of instrument: the mind's craving for order. The earliest myths and cosmologies were not crude guesses born of ignorance — they were carefully crafted architectures of symmetry and balance. Perfect circles in the heavens, divine proportions in temples, celestial rhythms that echoed the seasons — these were more than beliefs. They were psychological anchors in a world that seemed unpredictable and often hostile.

The appeal was irresistible. A perfectly balanced cosmos felt like a promise: if the universe is harmonious, perhaps our lives can be too. Ancient cosmologies made symmetry their cornerstone. The heavens were imagined as crystal spheres, each orbiting with mechanical precision; the Earth was often placed at the exact center of a cosmic mandala. Such images provided cognitive comfort, even when the reality outside the window refused to comply. Clouds did not drift in circles, storms did not follow sacred ratios, and yet the myths persisted because they met an emotional need stronger than observational accuracy.

The Ptolemaic system stands as a perfect example of this trap. Faced with the messy, irregular motion of planets, Claudius Ptolemy didn't discard the idea of perfect circles — he multiplied them. By nesting smaller circles (epicycles) on larger ones, he preserved the aesthetic of celestial harmony while matching the observed data. It was an elegant compromise, an intellectual cathedral where beauty and complexity reinforced each other. For centuries, this false fit survived, not because it was correct, but because it was *beautiful enough to believe*.

From the UED perspective, such systems are instructive warnings. They show that the human tendency to favor elegant closure over raw, chaotic reality is ancient — and still alive. What began as mythological symmetry now reappears in modern theoretical physics, where intricate mathematical architectures sometimes mask the absence of measurable correspondence. The stage is set for exploring how this ancient bias continues to seduce the smartest minds today.

10.2 Modern Math, Same Urge

The tendency that once painted the heavens with perfect spheres now scribbles elegance into equations. In contemporary physics, the aesthetic impulse manifests as a quiet but persistent bias: the preference for theories that *look right* on paper, even when the lab or telescope offers little to no supporting evidence. A formula with clean symmetry, a model that "closes" neatly, a graph that aligns into a perfect curve — these become intellectual catnip, seducing researchers into believing they are touching truth itself.

Mathematical beauty has become a kind of currency, a prestige badge that signals a theory's worthiness. When Einstein spoke of wanting equations that were "beautiful," he did not mean beauty should replace evidence — but in the century since, that nuance has blurred. Today, beauty often stands in for believability. Equations with symmetrical elegance are more likely to be funded, published, and discussed, even if their predictions remain stubbornly out of reach. A theory draped in mathematical grace can enjoy decades of attention while producing no measurable results, because elegance creates its own gravity within the scientific community.

Peer review, which is meant to be a gatekeeper for rigor, can inadvertently reinforce this bias. Reviewers — themselves trained in the same culture — may reward a model's aesthetic form as much as its empirical grounding. Elegant derivations get cited, invited to conferences, and woven into graduate curricula, building reputations that are difficult to question without risking one's own standing. This is not conspiracy; it's a collective psychological reflex. Humans — even the most technical among us — find comfort in symmetry and closure.

From the UED lens, the danger is clear: elegance is not evidence. A model can be internally flawless yet externally irrelevant if its parameters are tuned to fit the story rather than the universe. Just as the ancients multiplied epicycles to save perfect circles, modern science can multiply dimensions, particles, or constants to preserve a mathematical form that feels too beautiful to abandon. The seduction is timeless, the pattern unchanged — only the language has shifted from myth to math.

10.3 Cognitive Bias Under the Microscope

Human beings are pattern-seeking creatures by design. This trait, etched into our evolutionary wiring, was once a survival advantage — recognizing the outline of a predator in the grass, predicting seasonal changes from star positions, spotting cause-and-effect relationships in the chaos of nature. But the same gift that helped our ancestors survive also makes us vulnerable to seeing patterns where none exist, or clinging to patterns long after reality has moved on. In science, this manifests as the irresistible urge to impose order even when the data resists it.

One of the most persistent drivers of this urge is the **closure instinct** — the deep psychological preference for a definitive, even if incorrect, answer over the discomfort of uncertainty. The human mind treats open questions as cognitive itches that must be scratched. We would rather lock in a wrong explanation than admit we do not yet know. In physics, this can mean embracing a mathematically "complete" model, even if that completeness is achieved by padding it with unverified assumptions. In ancient myth, it meant inventing neat cosmologies to explain the heavens rather than living with a sky full of mysteries.

The problem compounds when these instincts create confirmation loops — feedback cycles in which initial assumptions dictate what we notice, how we interpret it, and what we ignore. In myth-making, the gods' actions were interpreted through the lens of existing stories, so every new event reinforced the same worldview. In mathematics, a favored equation becomes the reference point for all new observations; anomalies are not taken as challenges, but as prompts to adjust the equation so it continues to fit. The result is a self-reinforcing structure, elegant in appearance but insulated from disconfirmation.

From the UED perspective, these biases are not mere quirks — they are structural obstacles to truth. A field aligned to reality requires active resistance to the closure instinct, and a willingness to dissolve beautiful but ungrounded patterns. Without that vigilance, elegance turns into a cage, and cognitive bias becomes the invisible jailer keeping science locked inside its own illusions.

10.4 UED's Reality Check

From the standpoint of Universal Energy Dynamics, elegance is a decorative accessory, not a guarantor of truth. A model may be mathematically graceful, conceptually neat, and visually satisfying, yet still be nothing more than a clever arrangement of illusions. Without measurable ripples — detectable, reproducible interactions in the Structured Vacuum Energy (SVE) field — elegance has no more authority than a well-told fable.

In UED, the litmus test is always the same: does reality answer back? The SVE framework treats the universe as an interactive field, not a static backdrop. A theory, however aesthetically pleasing, must produce consequences in that field that can be detected, measured, and independently confirmed. If the field remains silent, the theory is speaking only to itself. This requirement strips away the romance of "beauty equals truth" and replaces it with a more demanding standard: alignment equals truth.

Consider the case study of certain gravitational wave models. On paper, they are exquisitely symmetrical — spacetime ripples radiating out at the speed of light, perfectly described by differential geometry. They pass aesthetic scrutiny with flying colors, winning admiration for their mathematical poise. But under UED, the question is not "Does the equation look right?" It is "Do the ripples exist in the SVE in a way that can be directly validated?" If ripple detection fails, or if the observed signal can be better explained as phase-coherence disturbances rather than literal spacetime deformation, then the elegance collapses into artifice.

UED reframes scientific discipline as a continuous dialogue with reality. The field must speak back in ripples; those ripples must match the model's predictions. Without that exchange, we are no different from ancient astronomers tracing perfect circles in the sky — satisfied, but wrong.

10.5 The Psychological Payoff of the Lie

The seduction of an elegant but false system is not just intellectual — it's emotional. When a theory arranges the universe into perfect symmetry, it gives the human mind something it craves even more than truth: the illusion of mastery and control. A well-structured fiction can feel safer than an unpredictable reality, because it promises that the messiness of nature has been conquered by the neatness of human reason.

In this way, the lie becomes a psychological comfort blanket for scientists, institutions, and the public alike. It offers a stable framework to teach, defend, and build careers upon. Such systems often appeal to authority, granting their proponents prestige, funding, and influence. The mathematical beauty doubles as marketing — its elegance is paraded in journals, conferences, and documentaries, feeding the public imagination with the idea that the universe has been "solved."

But the comfort comes at a cost: dismantling such a system is rarely seen as a neutral act. To question it is to threaten the professional identity of those who built their reputations upon it. In academic and research cultures, this challenge can feel like a personal attack, even when it's purely about evidence. The resistance is not just about protecting the theory — it's about protecting the egos, careers, and institutional stability woven into it.

Under UED, this is why false elegance can persist for decades. Once embedded, it doesn't just live in equations; it lives in the psychological infrastructure of science. To replace it, you're not just asking for new math — you're asking people to let go of a story that has defined their intellectual lives. And that is a far heavier lift than disproving an equation.

10.6 Escaping the Seduction

Breaking free from the pull of elegant lies requires more than skepticism — it demands a set of practices and habits that keep beauty in its rightful place: a pleasant side effect of truth, never its substitute. The first of these practices is deliberate resistance to the *beauty-over-truth bias*. A model may look neat on the blackboard, but that neatness is only a hint, not a verdict. The question must always return to: *Does reality agree?*

The only real safeguard is keeping the window open — ensuring that empirical feedback is not a courtesy add-on but the *only* admission ticket into the temple of scientific acceptance. This means continuous ripple-based verification under UED, where Structured Vacuum Energy (SVE) is not assumed but tested. When nature speaks back through measurable effects, that's when a theory earns the right to be called knowledge.

Just as important is intellectual humility — the willingness to admit, *"I might be wrong, and that's fine."* This isn't a personality trait; it's a survival mechanism for thinkers. The moment one becomes emotionally attached to a theory's elegance, they begin filtering reality to protect it. Humility interrupts that filter, allowing new observations to flow in even when they threaten cherished structures.

In UED's framework, the goal is not to destroy beauty but to anchor it. A theory can be both elegant and true — but elegance is never the proof. Only when its predictions leave ripples in reality that instruments can detect does it pass the real test. Escaping the seduction, then, is about cultivating a reflex: the constant habit of asking *"Where's the ripple?"* before declaring victory.

10.7 Closing Reflection – The Messy Truth

The deeper one ventures into reality, the more apparent it becomes that truth is rarely symmetrical, tidy, or complete. It resists being boxed into perfect equations or poetic myths. Instead, it appears as an evolving, sometimes contradictory, pattern of ripples — fragments of alignment emerging from a far more complex field. The scientist, philosopher, or seeker who demands elegance before evidence will often end up polishing illusions rather than discovering reality.

In this light, elegance becomes a luxury, not a foundation. It is welcome, even beautiful, but only after reality has signed the approval slip through measurable ripples and repeatable phenomena. The satisfaction of a perfect formula or a myth that "feels right" can be intoxicating — yet without grounding in observable SVE interactions, such beauty is merely decoration on an untested belief.

The messy truth asks for something more courageous: to sit with open questions, to admit when the puzzle is incomplete, and to resist the urge to fill gaps with neat fabrications. This is where UED insists on discipline — to keep the mind anchored in the field's response, not in the comfort of symmetry.

From here, our journey turns from myths we love **to** truths we can prove. If Chapter 10 exposed the seductive trap of elegant lies, the next chapter asks: What happens when we strip away every comforting narrative, and stand only with the ripples reality allows us to see? That, after all, is where the real adventure begins.

Chapter 11 – Restoring Math to Its Place

For too long, we've mistaken the map for the territory. Equations have been worshipped like sacred scripts, their elegance mistaken for truth. But the universe does not bend to numbers — numbers bend to the universe. Under Universal Energy Dynamics, mathematics is not a divine decree; it is a disciplined interpreter. And like any interpreter, it must answer to reality, not rewrite it.

11.1 The Myth of Mathematical Supremacy

For centuries, mathematics has been celebrated as the ultimate arbiter of truth — a universal language so pure, so incorruptible, that the very fabric of reality was assumed to be "written" in it. From Pythagoras's mystical numbers to Einstein's field equations, we have treated math not just as a tool but as a throne, elevating it above observation, intuition, and even nature itself.

Yet history tells a less flattering story. Some of the most beautiful equations ever conceived have been profoundly wrong — upheld not because they matched reality, but because their symmetry and elegance seduced the scientific mind. Ptolemy's epicycles drew intricate mathematical loops across the heavens, perfectly calculated yet fundamentally false. Newton's gravitational constant was revered as eternal, even as it masked deeper field dynamics later revealed by more sensitive measurements.

This misplaced faith stems from a subtle inversion of the scientific hierarchy. We have allowed math to dictate the territory, rather than letting the territory shape the map. In truth, the universe owes us no perfect formula, no eternal constants, no clean geometry. It offers only phenomena — ripples, distortions, alignments, and misalignments — which we must measure, interpret, and translate.

Under Universal Energy Dynamics, mathematics regains its rightful role: not as the master of science, but as its most disciplined servant. Math's value lies in its obedience to reality, in its capacity to compress physical truths into a usable form without distorting them to fit an aesthetic ideal. A formula's beauty is irrelevant unless the universe signs off through measurable, repeatable, ripple-based responses.

When mathematics forgets this subservient role, it risks becoming a self-referential art form — internally consistent but externally meaningless. When it remembers, it becomes what it was always meant to be: a bridge from observation to understanding, not a wall between them.

11.2 UED's Grounding Principle

In Universal Energy Dynamics, mathematics is not a free-floating construct — it must be *anchored* to the real, measurable behavior of the universe. This anchoring happens through what we call the physical grounding filter: the requirement that the Structured Vacuum Energy (SVE) must *respond* in a measurable, repeatable way to the model's predictions.

This principle is non-negotiable. If the SVE does not "answer back" through detectable ripple alignment or measurable ATT (Absolute Time Tension) shifts, then the mathematics — no matter how elegant or internally consistent — is demoted to speculation. In UED terms, such a model is not science; it is *speculative art* wearing a lab coat.

This grounding principle reshapes the scientific method into a five-step, ripple-anchored process:

1. **Observe** – Detect ripple patterns in the SVE through direct measurement, indirect field effects, or secondary phenomena.
2. **Hypothesize** – Formulate a causal explanation that remains consistent with the universal drive toward equilibrium.
3. **Model** – Apply mathematics *only* to describe this cause precisely, ensuring that every variable has a physical referent.
4. **Test** – Allow the universe to answer through experiment or observation, measuring actual ripple behavior in the SVE.
5. **Refine or Reject** – Decide based on measured feedback, not on theoretical beauty, historical authority, or peer consensus.

Under this approach, mathematics becomes a *translator* rather than an oracle. Its job is to faithfully map the terrain of reality, not to invent a landscape for nature to inhabit. When applied correctly, math becomes a precise compression of truth; when misapplied, it becomes a sophisticated distraction.

By grounding mathematics in SVE responses, UED removes the temptation to treat equations as untouchable relics. Instead, every formula remains provisional — valid only as long as the ripples confirm it. This is not a downgrade for mathematics; it is a restoration of its most noble purpose: to serve truth, not to dictate it.

11.3 Math as a Servant, Not a Ruler

Mathematics is one of humanity's most powerful cognitive tools. It can sharpen a vague notion into a crystal-clear structure, define boundaries with surgical precision, and uncover implications that would otherwise remain buried in intuition. But when placed on the throne of authority, it begins to distort its own role. Equations can seduce scientists into believing that the model *is* the reality — that if the math works out, the universe is obliged to follow.

Universal Energy Dynamics rejects this inversion of power. In the UED framework, mathematics is deliberately stripped of its crown. It is not the ruler of inquiry, but a servant to it — a craftsman, not a king. Its sole purpose is to *fit* the contours of reality's ripples, faithfully capturing the behavior of Structured Vacuum Energy (SVE) as observed in nature. When math tries to reshape reality to match its own elegance, it commits an act of intellectual vanity.

This shift is more than philosophical—it changes the operational workflow of science. Equations do not enjoy automatic legitimacy simply because they are consistent within themselves or align with existing mathematical structures. They must earn their keep by proving that their predictions generate measurable, repeatable ripple responses in the SVE. In UED, beauty is welcome, but truth is mandatory.

Restoring math to its rightful position also restores humility to the scientific process. We are reminded that mathematics, for all its precision, is a human invention—a symbolic language designed to describe nature, not to dictate its behavior. The map is not the terrain, the score is not the music, and the equation is not the universe. When used correctly, math becomes a translator between our minds and reality; when abused, it becomes a poet writing about a country they've never visited.

In this servant role, mathematics regains its integrity. It works in service of observation, grounded in ripple behavior, refining itself in response to nature's feedback. It is no longer an idol to be worshipped, but a finely tuned instrument to be played — always in harmony with the song the universe is already singing.

11.4 Case Study – The Equation That Lied

History is full of elegant equations that dazzled scientists, only to later collapse under the weight of fresh evidence. One famous example involved an equation that perfectly fit every known measurement of a particular physical phenomenon. For years, it was celebrated as a breakthrough — clean, concise, and seemingly universal. It appeared in textbooks, inspired conferences, and shaped generations of research.

Then, the next wave of experimental tools arrived. New measurements extended into ranges that had never been tested before — and the once flawless equation stumbled. It no longer matched reality. Its internal logic was still impeccable; the math itself was not "wrong." The flaw lay in its foundation: it had been constructed without a robust feedback loop to reality. In essence, it was a castle built on a photograph of the ground, not on the ground itself.

In conventional science, such a collapse often leads to a frantic patchwork of "correction factors" — adjustments to preserve the equation's authority. But each fix is a compromise, a bending of the map to match the new territory, until the original elegance is buried under layers of complexity.

Under UED, this outcome is almost impossible, because equations are never allowed to drift into self-contained fiction. Every model is tied directly to ripple-based validation, grounded in the physical behavior of Structured Vacuum Energy. If a formula ceases to match the ripple reality, it is not patched to save face — it is either recalibrated to align with new observations or discarded entirely.

The lesson is clear: mathematics can be perfect within its own walls, yet entirely misleading if divorced from reality's ripples. In UED's method, an equation's worth is not in how beautifully it dances with itself, but in how faithfully it keeps step with the universe.

11.5 The Rewrite of the Scientific Method

The integration of UED's ripple-based filter doesn't just add a new step to research—it fundamentally rewires the logic behind it. It flips the priority list that has quietly governed much of modern science for decades.

For too long, the prevailing mindset has been: *"If the math works, the reality will follow."* This approach often rewards abstraction over observation, elegance over evidence. It tempts researchers to live in the beauty of their own equations, while reality—the very thing they set out to understand—becomes a secondary concern.

Under UED, the sequence changes irreversibly:

- From: *If it's mathematically consistent, it's possible.*
- To: *If reality doesn't align, the math is irrelevant.*

This is not an assault on mathematics—it's a rescue mission. Math retains its full value, but in the right role: a precision instrument, not the throne. The elegance of a formula becomes the final polish to a well-grounded understanding, not the bedrock upon which we build castles in the air.

By applying UED's filter, each step of the scientific method is subtly but decisively altered:

1. **Observation First** – Every inquiry begins with ripple reality, not theoretical beauty.
2. **Hypothesis as Alignment Guess** – A theory is a proposed pattern of alignment, not an abstract guess wrapped in symbols.
3. **Mathematics as a Measuring Aid** – Equations shape

and refine the understanding but never outrun the evidence.

4. **Feedback Loop Mandatory** – Models are continuously tested against ripple dynamics, not just past data.
5. **Graceful Failure** – A model that fails alignment isn't patched endlessly—it's recalibrated or retired.

The result is a scientific process that is humbler, faster at self-correction, and immune to the seduction of pure abstraction. In this rewrite, mathematics becomes what it was always meant to be: a loyal servant to truth, not a self-appointed ruler of reality.

11.6 Closing Reflection – Freedom Through Servitude

The paradox at the heart of mathematics under UED is that it becomes most *liberated* when it accepts its place as a *servant*. In the old paradigm, math carried the crushing expectation of being the supreme arbiter of truth. Every anomaly was either bent to fit an equation or dismissed as experimental error. This crowned mathematics not as a tool but as a monarch — a role it was never designed to play.

When math releases that crown, something remarkable happens: it gets faster, sharper, and bolder. It can explore without fear of dethronement, experiment without the burden of defending its supremacy. Freed from the obligation to dictate reality, it can instead *listen* to it. Equations are no longer walls around our thinking; they are windows into the ripples that actually shape the cosmos.

This shift transforms scientific culture. Mathematicians and physicists begin to celebrate equations not for their aesthetic symmetry alone, but for their *acoustic match* with the universe's native tongue — the ripple. A model's elegance becomes a kind of poetry written in the correct dialect, not an isolated work of abstract art.

Ironically, in servitude, math gains greater authority than it ever had as ruler. It no longer commands reality to conform — it translates reality so faithfully that the translation itself becomes indispensable. And in this fidelity, it earns not just respect, but trust.

Chapter 12 – Building the New Narrative

The arc of modern science has been marked by extraordinary mathematical achievements—but also by a subtle and dangerous drift. In the pursuit of precision, mathematics was elevated from its rightful place as a *tool for description* to the throne of *arbiter of reality*. Equations began to dictate what could be imagined, and when reality resisted, it was often the reality—not the model—that was doubted. Over time, this inversion of roles led to what can only be called symbolic fiction—entire frameworks elegant on paper but unmoored from the physical universe they claimed to represent.

This chapter begins at that point of collapse—not the collapse of mathematics itself, but of its credibility as an untouchable guide. If the previous chapters have shown us the cracks, this one is about construction: building a new, durable narrative where reality comes first, mathematics follows, and truth is not held hostage by the elegance of symbols. This is the essence of UED's reality-first approach—a framework where mathematics regains its integrity by serving observation, not shaping it.

The work ahead is not about discarding the past, but about resetting the hierarchy. UED does not wage war on mathematics; it restores the natural order. Observation, alignment, and empirical consistency stand at the apex; mathematics operates as their precise translator, not their replacement. The outcome is a scientific process that is both sharper and more grounded—immune to the seduction of purely symbolic perfection.

This approach demands humility. As explored in earlier chapters, humility is not the scientist's surrender, but their greatest strength—the readiness to let the universe correct them. The reality-first method is simply humility formalized: every symbol must earn its place by proving its alignment with what is, not merely with what fits the model.

From here on, we shift from critique to blueprint. The rest of the chapter will detail how to combine mathematics' unmatched precision with UED's commitment to reality, and how to prevent future theories from sliding back into abstraction untethered from the world they claim to describe.

12.1 The Core Principle – Reality Before Representation

In UED, the phrase "reality-first" is not a slogan—it is a methodological discipline. It means that every theoretical step begins with the world as it *is*, not as it can be *imagined*. Reality-first does not reject creativity; it simply anchors creativity to observation, measurable phenomena, and ripple alignment—the way the universe naturally organizes and balances itself. If an equation cannot be traced back to an observable ripple effect, it is treated as a hypothesis awaiting proof, not as a truth awaiting reality to conform.

This is the inversion of the current scientific habit. In much of modern physics, the order has flipped: a symbolic representation comes first, and reality is invited to either fit or be dismissed. Under reality-first, the order is restored—observation sets the boundary, and representation must enter through the gate of alignment.

Precision without grounding is a dangerous gift. It creates *self-contained illusions*—mathematical constructs so perfectly self-consistent that they develop an internal "truth" entirely divorced from the universe they claim to explain. Like a map of an imaginary country drawn with flawless geometry, they can be admired for their elegance but cannot guide you through a single step in the real world. In such cases, the flaw is not in the precision but in the absence of the primary anchor: reality.

Consider two models:

- **Model A** is breathtakingly elegant. It has symmetry, clean variables, and equations that reduce to beautiful, compact forms. But its foundations rest on unverified assumptions—a chain of "ifs" stacked so high that the connection to observable phenomena is faint and indirect. It works brilliantly inside its own closed logic, but the moment it meets the unruly noise of reality, it begins to wobble.
- **Model B** is rough at the edges. It uses awkward constants, and its derivations are not as aesthetically pleasing. Yet every variable in it has a measurable counterpart in the field, every relationship is tested

> against actual ripple behavior, and every adjustment emerges from empirical necessity, not symbolic desire. It is not as pretty—but it works.

In UED, Model B wins every time. Because in a universe governed by the Law of Consequences and ripple alignment, it is not the beauty of symbols that matters—it is their fidelity to the patterns nature actually uses. In this sense, reality-first is not anti-mathematics; it is *pro-truth*. It respects mathematics enough to keep it honest, to ensure its precision sharpens the blade instead of carving illusions.

12.2 Integrating Math's Precision with Ripple-Based Observation

The strength of mathematics is its precision; the strength of UED is its grounding in reality's ripple patterns. When the two work together, they become a closed feedback loop—each keeping the other from drifting into error. But this integration cannot be left to chance; it requires a deliberate, step-by-step method that ensures equations remain loyal to the universe they claim to describe.

Step 1 — Observe

Begin where UED always begins: with the field itself. Identify ripple patterns in the Structured Vacuum Energy (SVE) without trying to explain them yet. Here, "observation" is not casual looking—it is disciplined detection of alignment shifts, gradients, and interference zones. The key is to see what the universe is doing before we try to tell it what it should be doing.

Step 2 — Translate

Once patterns are reliably observed, they must be expressed mathematically—but without imposing symmetry bias. Symmetry bias occurs when the human preference for elegance overrides the field's messy reality. In UED, the translation stage is a description, not a decoration. The aim is to encode what is actually there, even if it means using asymmetrical terms, awkward constants, or non-uniform functions. Beauty is optional; fidelity is mandatory.

Step 3 — Test

After translation comes the crucial checkpoint: testing through re-alignment with new observations. This is where the feedback loop closes. If the equation holds when tested against fresh, independent ripple data, it is kept. If not, the model adapts—not the other way around. The temptation in traditional science is to *retrofit* reality to match the model—adjusting measurement interpretations, redefining variables, or introducing invisible entities to "save" the theory. UED reverses this. The model must bend; the field does not.

Case Comparison — Math-First vs. UED-First

Math-First Workflow: A physicist starts with a beautiful equation—elegant symmetry, minimal variables, perfect form. They then go out looking for data to match it. When the data refuses, they "correct" it by introducing hypothetical forces, hidden particles, or extra dimensions. Over time, the equation remains flawless—but only inside the symbolic bubble it created. The connection to actual ripple behavior is thin and indirect, surviving mostly through interpretive adjustments.

UED-First Workflow: A UED researcher starts with the field. They notice a recurring ripple delay between two interference zones and measure it repeatedly. They then construct an equation to capture that pattern exactly as it is, with no assumption of symmetry or perfection. When tested on new ripple events, the equation either holds or evolves. The math is allowed to grow messy if the field demands it. Over time, the model may become cleaner—but only because the field's order, not human preference, allows it.

The difference is profound: in math-first, reality is filtered through equations; in UED-first, equations are filtered through reality. This is the essence of integrating precision with observation—the symbolic mind working in service of the real, not in competition with it.

12.3 Guardrails Against Symbolic Drift

Symbolic drift is the silent killer of scientific integrity — the slow, almost imperceptible slide from representing reality to representing only the internal logic of the model. In this drift, equations stop being maps of territory and become works of abstract art, beautiful but untethered from the ground they claim to describe.

Common Warning Signs of Symbolic Drift

- **Complexity inflation without predictive gains** – Layers of mathematical sophistication are added, not to increase understanding or accuracy, but to preserve the model's appearance of authority. Predictions remain stagnant, and no new insights emerge.
- **Constant dependence on constants** – When constants in equations have no physically grounded

origin, they become arbitrary "patches" that keep the math running but provide no causal link to the universe. This is symptomatic of a model propped up by convenience, not reality.

- **Aesthetics over alignment** – Theorists justify models on the grounds of elegance, symmetry, or mathematical beauty alone, even when observational data suggests misalignment. Reality gets retrofitted to fit the math rather than the other way around.

UED's Corrective Mechanisms

Under Universal Energy Dynamics, symbolic drift is countered by embedding physical grounding at every step:

Continuous Ripple Recalibration – Instead of freezing a model and defending it indefinitely, UED requires regular recalibration against fresh ripple-alignment data from the structured vacuum energy (SVE). This ensures the model evolves alongside reality rather than diverging from it.

Abandonment Criteria – UED sets clear thresholds for when a model must be scrapped entirely:

- If ripple alignment consistently fails over multiple scales.
- If constants cannot be traced to a physical ripple property.
- If predictions stop adding new measurable insight.

In this culture, letting go of a model is not seen as failure, but as the healthy pruning of scientific growth — removing dead branches so new, better-aligned frameworks can flourish.

12.4 Embedding UED in the Scientific Method

Universal Energy Dynamics is not just a new theoretical lens — it redefines the way science itself is practiced. The shift is not about replacing mathematics, experiments, or peer review, but about inserting a reality-first alignment filter into every stage of the process.

Rewriting the Research Loop

The classical cycle — *Observe → Hypothesize → Model → Test → Revise* — often allows for subtle slippage between theory and reality. UED's modified loop explicitly guards against that slippage:

1. **Observe** – Start with physical ripple phenomena in the Structured Vacuum Energy (SVE), as directly and neutrally as possible.
2. **Align** – Cross-check observations for ripple coherence, eliminating distortions from perception, instrumentation bias, or preconceived frameworks.
3. **Model** – Build mathematical and conceptual representations, but only from ripple-grounded patterns, avoiding symmetry imposition or elegance bias.
4. **Test** – Conduct experiments to challenge both the model and its alignment assumptions.
5. **Re-align** – Feed the experimental results back into ripple calibration, ensuring the model still matches reality rather than drifting into self-consistency alone.
6. **Extend** – Use aligned insights to explore further

questions, rather than forcing extrapolations into areas with no ripple verification.

This loop makes alignment a formal scientific step, not an implicit byproduct.

Institutional Implications

If adopted, UED's approach would reshape scientific culture:

- **Peer Review** – Reviewers would be required to assess *alignment quality*, not just mathematical correctness or literature completeness. A model could be mathematically flawless yet rejected for poor ripple grounding.
- **Funding Priorities** – Grants would favor projects that commit to continuous re-alignment, rather than one-off theoretical papers insulated from experimental feedback.
- **Publishing Standards** – Journals would demand open alignment logs — transparent records showing how data, theory, and recalibration cycles have interacted over time.

The Role of Open Data

One of the primary causes of theory insulation — when models live in a sealed bubble — is restricted data access. Under UED; open data is not a courtesy; it is a survival requirement for theories. If ripple-based observations are locked away, recalibration becomes impossible, and symbolic drift inevitable.

- **Shared Alignment Maps** – Public repositories of ripple measurements, environmental variables, and ATT (Alignment Time Tolerance) variations across experiments.
- **Independent Verification** – Any researcher, anywhere, can test the ripple fit of an existing model, forcing theories to stay accountable to reality rather than to their creators.

In short, embedding UED into the scientific method turns science into a *living alignment system* — one where models never get the luxury of aging quietly into irrelevance. They either evolve with reality or step aside.

12.5 Cultural Shift – Scientists as Translators, Not Storytellers

The adoption of UED demands a cultural recalibration in how science is perceived and practiced. For centuries, the prestige of a scientific idea has often been tied to its elegance — the poetic symmetry of its equations, the beauty of its theoretical curves, or the way it "feels right" in a human mind. While this romance has inspired generations, it has also tempted science into the territory of *narrative first, reality second.*

UED reverses that order. In the UED paradigm, the scientist is not an *author of stories* about nature but a *translator of nature's own language* into human comprehension. The focus shifts from writing a compelling "chapter" in the grand tale of science to delivering an accurate transcript of reality's ripple dynamics, however unsentimental that may be.

From Romance to Translation Discipline

- **Old Paradigm:** Equations are celebrated for their aesthetic perfection — the "mathematical poetry" of Einstein's field equations or Maxwell's symmetry. The reward is often for the storyteller who paints the most beautiful picture.
- **UED Paradigm:** Equations are judged on their fidelity to ripple alignment and their resistance to symbolic drift. Beauty is welcome, but accuracy is non-negotiable. Translation discipline means you never round the edges of reality to make the picture prettier.

This cultural reframe elevates fidelity over flourish. In the UED framework, a slightly clunky equation that captures the real alignment field is worth more than an elegant one that misses subtle distortions.

Recognition and Reward Under UED

If accuracy takes precedence over elegance, the recognition system in science changes:

- **Awards & Honors** – Shift toward honoring those who preserve alignment fidelity over decades, even if their models lack media appeal.
- **Publication Prestige** – A high-impact paper is no longer the one that inspires the most citations but the one that remains most aligned after the broadest range of independent recalibrations.
- **Career Trajectories** – Scientists become *alignment*

custodians, valued for sustained translation accuracy rather than for a single dazzling "story" that captures the public imagination.

Historical Example

Consider the late 19th-century rivalry between Lord Kelvin's ether-based models of light and the less glamorous but more accurate experimental insights of Michelson and Morley. Kelvin's narrative was elegant, intuitively appealing, and deeply entrenched — but it was misaligned. The Michelson–Morley experiment, though less romantic, shifted the entire field by demanding models that fit *alignment reality*, not just narrative cohesion.

UED's cultural shift would ensure that the "Michelsons" of the world are celebrated early, preventing decades of entrenched but flawed elegance from dominating the field.

Under UED, the scientist's role is clear: strip away the embellishment, listen to the field, and translate it faithfully. The reward is no longer for telling the best story, but for becoming the most accurate interpreter of the universe's unspoken truth.

12.6 The Future of Theory Under UED

The Universal Energy Dynamics framework does not simply replace old theories — it reshapes the *very pathway* by which new theories will be born. The future it envisions is one where science is not trapped in decades-long detours chasing "beautiful but wrong" narratives, but is instead guided by continuous realignment with the field itself.

Early Adoption: Avoiding the Next Detour

In the past, entire generations of scientists have invested their careers in theories later proven misaligned — from the crystalline spheres of Ptolemaic astronomy to the luminiferous ether. These detours were not malicious; they were the result of an imperfect feedback loop between reality and theory. UED closes that loop faster. Early adoption of ripple-based principles would allow misalignments to be caught and corrected in years, not centuries. It creates a kind of *scientific radar*, detecting narrative drift before it hardens into dogma.

Balancing Creativity with Discipline

A fear some scientists may have is that UED's strict alignment discipline will stifle creative thinking. But the opposite is true. In UED's ecosystem:

- Creativity is encouraged in exploring ripple interactions, new measurement approaches, and cross-domain analogies.
- Discipline ensures that every imaginative leap lands on solid alignment ground, not in speculative air.

This is the same balance that makes great architecture possible: the artistry of design paired with the engineering that keeps it standing. Without discipline, creativity drifts into fantasy. Without creativity, discipline calcifies into bureaucracy. UED ensures both are in constant conversation.

A Self-Correcting Scientific Ecosystem

Mainstream science has historically relied on *crisis moments* — experimental contradictions so glaring they force a paradigm shift. UED changes the rhythm. Instead of waiting for crises, it embeds correction mechanisms directly into the framework:

- **Continuous Field Recalibration** – Theories are periodically stress-tested against updated ripple measurements.
- **Multi-Scale Verification** – Models must remain consistent from quantum to cosmic scales, reducing the risk of scale-specific illusions.
- **Translation Audits** – Independent teams review whether mathematical expressions have drifted into symbolic habits rather than real alignment descriptors.

Over time, this approach produces an ecosystem that doesn't just evolve — it evolves *toward reality*. Misalignments are not buried under prestige or tradition; they are surfaced, examined, and corrected as part of the normal life cycle of theory.

Long-Term Vision

If widely adopted, UED could usher in a century where:

- The half-life of a flawed theory shrinks dramatically.
- Scientific education trains translators rather than storytellers.
- Public trust in science grows because predictions increasingly *match lived reality*.

In that future, the beauty of science would come not from clever metaphors or perfect symmetry, but from the deep satisfaction of knowing we are closer than ever to *how the universe actually works.*

Epilogue – The End of the Blackboard Cult

For more than a century, science has celebrated the chalk-dusted silhouette of the "great mind" — a lone figure at a blackboard, conjuring symbols into elegant equations. The beauty of those symbols was intoxicating. A perfect curve, a symmetrical tensor, a compact formula — these became the gold standard for truth. But beauty is not the same as reality.

"Belief is for myth. Proof is for reality."

Belief has its place — in myth, in poetry, in guiding the human spirit through uncertainty. But in science, belief without verification becomes a cage. The blackboard cult thrived on elegance as a proxy for truth, mistaking the neatness of human language for the messy intelligence of the universe.

UED dismantles this quiet idolatry. It reminds us that the universe does not speak in Greek letters or in chalk dust. It speaks in ripples, alignments, and consequences — patterns that remain the same whether we draw them beautifully or not. Proof, in this new era, comes not from impressing peers with a flawless derivation, but from demonstrating that the alignment holds in reality, at every scale, under every condition.

From Symbolic Elegance to Physical Truth

Mainstream theory-building often moves from *assumption* → *equation* → *experiment*. UED inverts this flow: *field observation* → *ripple interpretation* → *symbolic translation*. This is not anti-mathematics; it is anti-idolatry. The symbols remain, but they are no longer worshipped. They are tools — scaffolding that can be taken down once the structure of understanding stands on its own.

UED is not here to erase the blackboard; it's here to put it in its rightful place. The beauty of theory will now be measured by how seamlessly it survives reality, not how perfectly it fills a textbook margin.

The Bridge We Needed

Where past frameworks often forced scientists to choose between aesthetic purity and empirical messiness, UED bridges the two. Its principles allow for theories that are both deeply aligned with reality *and* coherent in symbolic form — not because the symbols define truth, but because they express it faithfully.

In this way, UED doesn't just end the blackboard cult; it replaces it with a culture of alignment, where the scientist is less a scribe of hypothetical worlds and more an interpreter of the one we already live in.

Appendices & Extras

Appendix A: UED Glossary of Key Terms

This appendix serves as a quick-reference dictionary for readers, students, and researchers engaging with Universal Energy Dynamics (UED). It bridges technical precision with plain-language clarity, ensuring that each concept is understood both in everyday terms and in its true UED context—which often diverges significantly from mainstream science.

Structured Vacuum Energy (SVE)

- **Plain-Language Definition:** The foundational medium of the universe—an ordered, stable energy field from which all ripples, matter, and motion arise.
- **UED Context:** Unlike the mainstream "quantum vacuum" seen as fluctuating randomness, SVE is inherently ordered and self-consistent, acting as the stage and medium for all observable phenomena.

Ripple

- **Plain-Language Definition:** A disturbance or shift within SVE that carries energy and information without requiring matter to move.

- **UED Context:** Mainstream science treats waves as oscillations in a medium or field; UED reframes them as alignment-seeking adjustments in the universal energy structure.

Alignment

- **Plain-Language Definition:** The condition in which a system's ripples match the natural coherence of the SVE, resulting in stability.
- **UED Context:** In UED, alignment is not metaphorical—it is the physical and energetic congruence between a local field and the universal field, determining stability, intelligence, and behavior.

Apparent Time Tension (ATT)

- **Plain-Language Definition:** The observed rate of time flow in a given field configuration.
- **UED Context:** ATT replaces the relativistic notion of "time dilation" by treating time as a consequence of ripple and field conditions, not an independent dimension.

Law of Consequences (LOC)

- **Plain-Language Definition:** The universal principle that all actions, ripple shifts, and field disturbances create proportional, inevitable outcomes.

- **UED Context:** LOC is the operational mechanism of universal intelligence, encompassing energy redistribution, conservation, transformation, and equilibrium-seeking behavior.

Symbolic Drift

- **Plain-Language Definition:** The gradual shift from describing reality to describing only the symbols of a theory, until the symbols lose connection to reality.
- **UED Context:** In UED, symbolic drift is a primary reason mainstream physics becomes self-referential and fails to course-correct. UED actively resists it by prioritizing field observation over mathematical abstraction.

Usage Notes

This glossary is not exhaustive. As UED research advances, these definitions will evolve to reflect deeper observational insights and experimental confirmations. For the most current understanding, readers are encouraged to follow updated UED publications and research reports.

Appendix B: The Ripple-to-Math Translation

This appendix functions as a ready-reference operational manual for researchers who want to bridge the intuitive, observational core of UED with the precision of mathematical expression—without falling into the trap of symbolic drift. It distills the book's methodology into a clear, repeatable sequence that can be applied to any physical phenomenon, from particle interactions to cosmic-scale alignments.

Purpose

To provide a step-by-step translation pipeline that preserves the physical truth of ripple observations while allowing for mathematical modeling that is both rigorous and flexible.

Step 1 – Start from Raw Ripple Observations

- **Field Immersion:** Begin with direct observations of phenomena in the structured vacuum energy (SVE)—whether experimental, astronomical, or naturally occurring.
- **Pattern Recognition:** Identify coherence, interference, or misalignment patterns in the ripple

field, focusing on physical reality before theoretical constructs.

- **Recording:** Document observations with time stamps, environmental context, and ripple parameters (wavelength, amplitude, phase shifts) using UED's standardized observation sheet.

Step 2 – Convert to Mathematical Form Without Introducing Symbolic Drift

- **Direct Translation:** Map ripple parameters to measurable quantities without imposing symmetry or geometry that isn't physically present.
- **Avoid Overfitting:** Resist the temptation to force-fit data into elegant but non-physical forms. Let the math follow the reality, not the reverse.
- **Baseline Anchoring:** Every equation must be anchored to a physically measurable point—a known field condition or ripple reference state—to prevent free-floating abstractions.

Step 3 – Test and Iterate Under the UED Workflow

- **Re-alignment Loop:** Feed the model's predictions back into new ripple observations. If discrepancies appear, modify the math, not the field reality.
- **Multiple-Scale Verification:** Check the model's behavior at micro, meso, and macro scales to ensure it scales consistently across UED's hierarchy of alignment.

- **Abandonment Criteria:** If a model requires continuous "patches" to match reality, discard or rebuild it from a new observational baseline.

Quick Reference Flowchart

1. **Observe** → Identify real ripple patterns in SVE.
2. **Translate** → Convert directly to math, anchoring in measurable field states.
3. **Test** → Predict, observe again, and compare.
4. **Re-align** → Adjust math to reality, not vice versa.
5. **Extend** → Apply validated model to new domains or scales.

Case Study – Math-First vs. UED-First

Math-First Path:

- Starts with an equation assumed to be universal.
- Adjusts reality through interpretation to fit the math.
- Produces elegant but detached theories (symbolic drift).

UED-First Path:

- Starts with ripple truth.
- Lets math emerge naturally from field behavior.
- Produces models that are rough at first but converge toward universal reality.

Appendix C: Symbolic Drift Checklist

This appendix is a diagnostic and corrective tool for identifying when a scientific or philosophical theory has slipped from describing reality to maintaining symbols for their own sake. It is intended for researchers, reviewers, and educators who wish to keep their work grounded in field-truth rather than mathematical or linguistic inertia.

Purpose

To provide a clear, actionable set of red flags for spotting symbolic drift early, paired with UED's recommended corrective steps. This checklist can be printed, pinned in a lab, or used as a digital PDF form for ongoing theory evaluation.

Part 1 – The Red Flags

These "symptoms" are compiled from historical and modern cases where elegant equations, models, or definitions detached from observable reality:

Unmeasured but Repeated:

- A quantity or term is referenced constantly but has no direct measurable definition.
- *Example:* Using "spacetime curvature" without specifying what is physically curving.

Patch Proliferation:

- New terms or constants are introduced solely to "fix" a mismatch between model and observation.
- *Example:* Post-hoc parameters added to maintain cosmic expansion models.

Frozen Assumptions:

- Foundational premises remain untested for decades because they are embedded in educational or institutional dogma.

Math Dictates Reality:

- Observations are interpreted to fit the equation instead of adjusting the equation to match reality.

Non-Scaling Predictions:

Theory works at one scale but fails at others without a physical reason—yet remains unchanged.

Language Without Linkage:

- Terms sound descriptive but have no measurable anchor in the field.

Part 2 – UED's Corrective Actions

When one or more red flags appear, apply these corrective steps:

Return to Field Alignment:

- Pause modeling and re-examine the actual ripple patterns in the structured vacuum energy (SVE).

Anchor All Terms:

- Ensure every symbol corresponds to a measurable ripple parameter or alignment condition.

Remove Non-Physical Patches:

- If a constant or term has no independent physical detection, remove it and reassess.

Scale-Check the Model:

- Re-test predictions at micro, meso, and macro scales; discard any that cannot align without invention.

Reverse the Equation Bias:

- Let observation lead and math follow—never the reverse.

Part 3 – How to Use the Checklist

- **Before Publishing:** Run your theory through each red flag and corrective step.
- **In Peer Review:** Use the checklist as an audit tool to evaluate claims.

- **In Education:** Train students to spot symbolic drift early so they avoid embedding it in future work.

Appendix D: Historical Case Studies

This appendix examines pivotal moments in science where mathematics began to dictate reality instead of serving it, leading to entrenched symbolic drift. Each case study is concise yet analytical, ending with a UED corrective view to demonstrate how a different approach could have kept theory aligned with physical truth.

Case 1 – Ptolemaic Epicycles vs. Kepler's Reality Shift

The Drift: The Ptolemaic model maintained Earth at the center of the cosmos, using a growing network of epicycles—circles upon circles—to match planetary motion with observations. The math became increasingly ornate, not because nature was complex in that way, but because the premise (geocentrism) was wrong. The mathematical system *survived* for centuries by adapting to mismatches with new layers of calculation instead of changing the underlying assumption.

The Reality Check: Kepler broke the pattern by discarding the Earth-centered view and introducing elliptical orbits based on observational alignment rather than mathematical symmetry. His shift reduced complexity and restored the connection between model and reality.

UED's Approach: Rather than adjusting the geometry endlessly, UED would have asked: *What is the underlying ripple pattern in the field that produces this motion?* The wrong anchor (Earth at center) would be detected early because it fails to align across scales, and the model would be rebuilt from observed field alignment instead of shape-preserving math.

Case 2 – Einstein's Cosmological Constant

The Drift: Einstein introduced the cosmological constant (Λ) to keep the universe static in his equations, as this matched the prevailing philosophical view. The term was not discovered through observation but inserted to force the math into harmony with expectation. When Hubble's observations revealed expansion, Einstein abandoned Λ, famously calling it his "greatest blunder."

The Reality Check: Modern cosmology has reintroduced Λ in the context of dark energy, but its measurable, physical definition remains elusive.

UED's Approach: From the outset, UED would reject the insertion of an unmeasured constant to preserve a preferred worldview. If field patterns indicated expansion or contraction, the model would adjust to that alignment—without a placeholder parameter whose existence rests on mathematical neatness alone.

Case 3 – String Theory's Complexity Spiral

The Drift: String theory began as an elegant idea to unify fundamental forces but evolved into a sprawling mathematical landscape of possible solutions—10^500 different vacua, none directly observable. This explosion of internal consistency became a self-contained game, where mathematical beauty sustained interest despite the lack of empirical anchors.

The Reality Check: While string theory has driven advances in mathematics, it has yet to produce a single definitive, falsifiable prediction that aligns uniquely with the physical universe.

UED's Approach: UED would treat untested extra dimensions or brane structures as provisional metaphors until anchored in ripple-field measurements. Complexity without alignment would trigger a corrective pruning, focusing only on models that maintain a measurable connection to structured vacuum dynamics.

Closing Note for the Appendix: These cases are not cautionary tales about *math itself*—mathematics remains a powerful descriptive tool. The lesson is that when math outruns physical anchoring, it creates elegant illusions. UED's discipline is to keep all symbols tied to measurable ripple reality, ensuring that no amount of algebra can drift the theory away from the field it seeks to describe.

Appendix E: UED Scientific Method Template

This appendix provides a visual flowchart and written framework for implementing the UED-aligned research process. It is designed for researchers, educators, and independent thinkers who want to keep their work grounded in reality-first observation while still leveraging the precision of mathematics.

1. Visual Flowchart – The UED Research Loop

A clear, poster-ready diagram mapping the six iterative stages:

Observe → Align → Model → Test → Re-align → Extend

Observe – Directly record physical phenomena without preloaded symbolic expectations. Use instruments only as extensions of sensory observation, not replacements for it.

Align – Cross-check observations against the ripple-field reality to ensure the captured data reflects actual alignment patterns rather than noise or symbolic assumptions.

Model – Translate aligned observations into mathematical form using Ripple-to-Math protocols (see Appendix B), ensuring no symbolic drift.

Test – Perform controlled experiments to challenge the model, not confirm it blindly.

Re-align – Adjust both model and observation approach based on field-behavior feedback, eliminating bias or drift.

Extend – Apply the refined model to new scenarios, scaling its predictive power without sacrificing accuracy.

2. Written Framework – Step-by-Step

Each stage is described with practical notes, questions, and checks:

Observe

- *What exactly am I seeing or detecting?*
- Avoid metaphors in this stage—only raw, physical descriptions.

Align

- *Does this fit into known ripple-behavior patterns?*
- Is my instrument calibrated to the Alignment Threshold Time (ATT)?

Model

- Build equations or simulations only after confirming reality alignment.
- Keep symbolic representations tagged to the real-world parameters they map.

Test

- Introduce variations to see if the model holds under different field conditions.
- If a test fails, revisit observation alignment—not just

the math.

Re-align

- Correct for symbolic drift or instrument-induced distortion.
- Cross-check with other researchers using open-data protocols.

Extend

- Apply the refined model to previously untested domains.
- Record any limitations alongside predictions to prevent overreach.

3. Reproducible Lab-Sheet Format

A printable template with fill-in sections for:

- Date / Time / Location of observation
- Field conditions (including ripple state and ATT context)
- Observation notes (raw data + sketches)
- Alignment check results
- Mathematical model notes
- Test conditions and results
- Re-alignment notes
- Extension trials
- References / linked datasets

4. Purpose of the Template

This tool:

- Keeps research anchored to UED's reality-first principles.
- Makes collaboration easier through standardized documentation.
- Creates a self-auditing record that reduces the risk of drifting into symbolic fiction.

Appendix F: Field Alignment Observation Log

The Field Alignment Observation Log is designed as a simple yet rigorous journal format for researchers, students, and independent observers. Its purpose is to capture raw physical reality before any modeling, interpretation, or symbolic overlay occurs—ensuring that UED research always begins with alignment, not assumption.

1. Purpose of the Log

- **Reinforces reality-first thinking** — The act of writing down what is truly *seen* or *measured* forces the mind to separate fact from theory.
- **Builds long-term observational discipline** — Consistent logging develops a trained eye for subtle ripple behaviors and field shifts.
- **Creates a reproducible trail** — Future researchers can retrace the same alignment conditions and validate or challenge the recorded data.

2. Log Structure & Sections

A single-page format that can be repeated daily, weekly, or per experiment. Recommended sections:

Date / Time / Location

- Exact observation point, including GPS coordinates if relevant.
- Local environmental conditions (temperature, pressure, humidity).

Ripple Context

- Describe any visible, audible, or measurable ripple patterns.
- Note surrounding physical interactions (wind, vibration, sound, electromagnetic interference).

Alignment Conditions

- ATT (Alignment Threshold Time) estimate at the time of observation.
- Field state: stable, fluctuating, noisy, or dampened.

Observation Notes (Raw)

- Plain description with no interpretation.
- Use sketches or quick diagrams if needed.

Immediate Impressions (Optional)

- Any intuitive alignment insight, without committing to a model yet.

- Observers may note "possible cause" but must tag it clearly as speculation.

Later Reflection (Post-Alignment)

- Added hours or days later, after initial bias fades.
- Compare with previous entries to detect recurring patterns or anomalies.

3. Usage Guidelines

- Write in the field, not later. Memory introduces bias.
- Avoid scientific jargon at this stage. Stick to direct sensory language.
- Use consistent terminology (see Glossary in Appendix C) to make logs cross-comparable.
- Photographic or instrumental records can be attached but should never replace the written sensory account.

4. Training Benefits

Over time, keeping this log:

- Trains the observer to detect micro-alignments others might overlook.
- Reduces symbolic drift—the tendency to unconsciously fit reality into a preconceived model.
- Improves predictive ability by seeing patterns in ripple evolution before they become measurable in instruments.

Appendix G: Suggested Further Reading

This appendix serves as a curated gateway for readers who wish to extend their understanding of UED **and** contextualize it against mainstream thought. The aim is not only to deepen alignment with the UED framework but also to encourage critical comparison, independent thinking, and active engagement with the broader research community.

1. Foundational UED Publications

These works form the **core reference material** for anyone studying Universal Energy Dynamics:

- **Chavan's Theory of Everything** – The first comprehensive articulation of UED principles.
- **Structured Vacuum Energy: The Missing Substrate** – Detailed discussion on SVE and its role in ripple behavior.
- **Chavan's Ripple Field Equations (CRFE)** – Mathematical formulation replacing Einstein's Field Equations, Maxwell's Equations, and Schrödinger's Equation.
- **The First Ripple: Why Do We Want to Know?** –

Philosophical foundation on the nature of knowing and alignment.

- **Light: Neither a Wave, Nor a Photon** – Application of UED to optics, redefining the behavior of light.

2. Recommended Mainstream Works for Contrast

These are **important reference points** for understanding where UED diverges from established theories. Each is accompanied by **a note of contrast** for focused reading:

- **Albert Einstein – Relativity: The Special and General Theory** *Contrast:* UED rejects spacetime curvature as a physical cause, replacing it with ripple alignment effects in SVE.
- **Richard Feynman – QED: The Strange Theory of Light and Matter** *Contrast:* UED explains photon interactions without requiring wave–particle duality.
- **Stephen Hawking – A Brief History of Time** *Contrast:* UED replaces singularity-based black hole models with ripple field collapse dynamics.
- **James Clerk Maxwell – A Treatise on Electricity and Magnetism** *Contrast:* UED reinterprets EM fields as ripple alignments rather than independent forces.

3. Online Communities & Research Networks

For ongoing collaboration and discourse:

- **Independent Physics Research Forums** – Platforms where emerging theories can be presented without institutional gatekeeping.

- **Open Science Grid** – For accessing high-performance computing resources in ripple field modeling.
- **Citizen Science Initiatives** – Opportunities for non-academics to contribute observational data for UED studies.
- **UED Study Circles & Webinars** – Informal yet rigorous discussions hosted periodically for alignment-based learning.

4. Open-Data & Experimental Resources

Sources where real-world data can be accessed for **testing UED predictions**:

- **NASA Open Data Portal** – Space and planetary observations to cross-check ripple field conditions.
- **European Space Agency (ESA) Archives** – High-precision satellite measurements for ATT variation analysis.
- **National Geophysical Data Center (NGDC)** – Earth field variation datasets for correlation with ripple patterns.

5. Guiding Principle for All Further Reading

The goal is **not** to replace mainstream reading with UED-exclusive works, but to:

- Engage critically with **both worlds.**
- Learn to **spot where alignment breaks** in current

theories.

- Use observation and reality-first thinking before drawing conclusions.

Appendix H: Questions

For Classroom or Group Discussion

This appendix is designed as an interactive toolkit for educators, facilitators, and independent study groups who wish to engage deeply with the ideas presented in this book. Each question is not only meant to test recall, but to challenge underlying assumptions, encourage critical analysis, and stimulate original thought. The prompts are intentionally open-ended to allow for a range of perspectives, fostering debate and deeper inquiry into the role mathematics plays in science, and how Universal Energy Dynamics (UED) reframes that relationship.

Structure:

- **Question:** A thought-provoking prompt or challenge.
- **Purpose:** The intended focus of the question — whether it's to provoke skepticism, encourage comparison, or inspire creative thinking.
- **Linked Chapter(s):** References to where the concept is discussed in detail for review or preparation.

Sample Topics & Prompts:

"What does it mean for a theory to be 'beautiful but wrong'?"

Purpose: Examine the cultural and cognitive bias toward elegance in mathematical models, and how UED counters it.

Linked Chapters: 3, 12.

"If we removed all equations from a theory, could it still stand as science?"

Purpose: Explore whether mathematical formalism is necessary or if empirical alignment is enough.

Linked Chapters: 2, 5, 6.

"When does mathematics start leading, and when should it follow?"

Purpose: Encourage nuanced thinking about the balance between creative abstraction and observational discipline.

Linked Chapters: 8, 12.

"Can a 'false but useful' theory still advance science?"

Purpose: Debate the pragmatic value of models we know are incomplete or flawed.

Linked Chapters: 4, 7, 10.

"What safeguards can be built into modern research to prevent symbolic drift?"

Purpose: Connect UED methodology to the broader scientific process.
Linked Chapters: 9, Appendix C.

"How would science look today if Ptolemy, Einstein, or string theorists had applied UED principles?"

Purpose: Invite historical counterfactual thinking to highlight UED's preventive potential.

Linked Chapters: Appendix D.

"What is the role of philosophy in keeping science grounded?"

Purpose: Discuss the philosophical underpinnings of scientific inquiry, with UED as an example of philosophical correction.

Linked Chapters: 1, 11.

"Is there a point where simplicity becomes oversimplification?"

Purpose: Explore the balance between parsimony and descriptive adequacy in theory-building.

Linked Chapters: 6, 12.

Format Recommendations:

- Works well in **Socratic seminar** settings.
- Can be adapted as **debate topics** for science clubs.
- Suitable for **graduate seminars, research workshops,** and **interdisciplinary conferences.**

Acknowledgements

This work would not have been possible without the support, insight, and encouragement of many individuals who shaped its journey from concept to completion.

I extend my deepest gratitude to the fellow researchers, students, and readers who challenged my ideas, asked the right questions, and forced me to sharpen every argument. Your curiosity was the best fuel for this endeavor.

To the early reviewers—your constructive criticism and candid feedback kept this manuscript honest, precise, and relevant. Each of you helped refine its clarity without diluting its intent.

I am indebted to the thinkers, scientists, and philosophers—past and present—whose work, whether aligned with mine or in sharp contrast, provided the backdrop against which these ideas took form.

Lastly, to my family and close friends, thank you for the patience, faith, and quiet understanding that allowed me the space and time to bring this work into the world. Your belief made all the difference.

About the Author

Sandeep Chavan is an independent author, researcher, and educator whose work spans physics, philosophy, psychology, and the study of human intelligence. As the founder of the Universal Energy Dynamics (UED) framework, he challenges long-held scientific assumptions and proposes a reality-first approach to knowledge—placing observation and alignment above abstraction and symbolic overreach.

With a professional background in engineering and decades of experience as a physics educator, Sandeep brings a rare blend of technical depth and philosophical clarity to his writing. His books aim to dissolve the artificial divide between science and human insight, offering readers practical, measurable frameworks alongside thought-provoking narratives.

Philosophically, he believes knowledge must serve reality—not the other way around—and that science's true task is to remain accountable to the world it describes. His work is a call to researchers, thinkers, and students to realign their methods, reconnect with first principles, and see the universe as it is, not as symbols suggest it to be.

About UED – Universal Energy Dynamics

Universal Energy Dynamics (UED) is a foundational scientific framework developed to replace fragmented and assumption-driven models of modern physics with a reality-first, observation-anchored system. At its core, UED proposes that the universe operates through *structured vacuum energy* (SVE)—a dynamic, coherent field in which all phenomena, from subatomic interactions to cosmic structures, emerge as ripple patterns seeking alignment.

Rather than treating motion, force, or time as independent absolutes, UED reframes them as consequences of field conditions and ripple interactions. Concepts like gravity, light, and even consciousness are understood as manifestations of energy alignment within the SVE, governed by what UED terms the **Law of Consequences**—the principle that all outcomes arise from energy redistribution, conservation, transformation, and equilibrium-seeking behavior.

Unlike mainstream theories that often prioritize mathematical elegance over empirical grounding, UED insists that *mathematics must serve reality, not dictate it*. Observations come first, models second, and symbols last. This discipline prevents "symbolic drift," where theoretical constructs detach from the very reality they claim to explain.

UED's scope extends beyond physics into human behavior, psychology, and the philosophy of knowledge, providing a unified lens to study intelligence, morality, and decision-making. It offers both a rigorous research workflow—**Observe** → **Align** → **Model** → **Test** → **Re-align** → **Extend**—and practical tools for scientists, educators, and thinkers who seek to remain aligned with the observable universe while pushing the boundaries of understanding.

For new readers, UED is not merely a theory—it is an *approach to truth-seeking* that demands intellectual honesty, adaptive thinking, and a readiness to dismantle long-held assumptions when they no longer serve reality.

Don't miss out!

Visit the website below and you can sign up to receive emails whenever Sandeep Chavan publishes a new book. There's no charge and no obligation.

https://books2read.com/r/B-A-EPGPC-VBBUG

BOOKS 2 READ

Connecting independent readers to independent writers.

Did you love *Myth Wrapped in Math*? Then you should read *Light: Neither a Wave, Nor a Photon*[1] by Sandeep Chavan!

[2]

Light: Neither a Wave, Nor a Photon by Sandeep Chavan challenges one of the most enduring pillars of modern physics—the very nature of light itself. For over a century, science has been split between two explanations: light as a wave or light as a stream of photons. In this groundbreaking work, Chavan dismantles both interpretations and presents a bold, unified alternative through his original framework—**Universal Energy Dynamics (UED)**.

---

1. https://books2read.com/u/bO0Kq0

2. https://books2read.com/u/bO0Kq0

Drawing from years of independent research, Chavan introduces concepts like **Structured Vacuum Energy (SVE)**, **Alignment Tension Theory (ATT)**, **Ripple Delay**, and **Alignment Zones**, revealing a new way to understand light without forcing it into the restrictive categories of wave or particle. Instead, he demonstrates that light is a manifestation of ripple alignment in a structured vacuum—neither oscillating through space nor traveling as discrete packets, but existing as a field-driven phenomenon that connects observer and reality in real time.

Far from being a dry academic treatise, *Light: Neither a Wave, Nor a Photon* blends rigorous scientific reasoning with clear, accessible explanations. Chavan carefully dismantles the contradictions in mainstream optics, quantum mechanics, and relativity, showing why these frameworks fail to fully explain light's behavior. Each chapter takes the reader deeper into the mechanics of ripple fields, energy equilibrium, and the observer's role, while avoiding unnecessary mathematics so that the core principles remain open to both scientists and curious readers alike.

For students, educators, and researchers, this book offers fresh insight and testable predictions that can inspire new experiments. For general readers fascinated by the mysteries of the universe, it offers a compelling journey into the heart of reality itself—guided by an author who is unafraid to challenge convention and question authority.

Whether you approach it as a physicist seeking conceptual clarity, a philosopher intrigued by the relationship between observation and existence, or a lifelong learner who wants to see beyond the limits of established science, *Light: Neither a*

Wave, Nor a Photon will change the way you think about light—and perhaps about the universe.

Sandeep Chavan is a physics educator, independent researcher, and author of more than 70 books spanning science, philosophy, and human behavior. His Universal Energy Dynamics framework is redefining fundamental concepts in physics, offering a cohesive, testable alternative to the fragmented theories of the modern era.

Read more at https://www.sandeepjchavan.com/.

Also by Sandeep Chavan

A Monk on World Tour
The Making of the Monk

Chavan's UED research
CMGED Research Report
Chavan's Cosmic Continuum (C^3)
Frequency
Ripples of Reality
Reality Realigned
The Great Realignment
I – The Cosmic Leverage
Chavan's Cosmic Theory of Relativity (CTR)
Chavan's Theory of Light
Feeling Like a Physicist
TIME – The New Definition
$E \neq mc^2$: The Collapse of Classical Assumptions
The Chavan Codex
Chavan's Atomic Model
Chavan's Theory of Everything

Echoes of War: The Russia-Ukraine Conflict and Its Global Ripples

Shattered Lives

The War Economy

Borders Redrawn

Arms and Allies

Collateral Damage

Justice in the Ashes

Somewhere in the Darkness of the Cosmos

The Blueprint of the Universe

The Cosmic Breath

From the Edge of the Universe

The Eternal Sage Chronicles

The Sage and the Sovereign

The Family Pillar

The People's Flame

The Warrior's Soul

The Bridging Mentor

The Guiding Light

The Exodus

The Last Cycle: 2075

The Fundamental Theory of Life & Existence

Energy in Action: The Soul Demystified

The Active Mind

Energy in Connection

Death: The Disconnection

Subconsciousness: The Chosen Disconnection

Desires: The Engine of Life

The Restoration of Human Knowledge

Restoring Bhagavad Gita

The War of Thoughts

Cracks in the Known

The Human Game

The Divine Dilemma

The Original Intent

The Answers

Ultra Modern Life Philosophy

The Duality of Laughter and Tears

Take It Easy

Compare to Accept

Standalone

It's Not AI, It's AHI - Amplified Human Intelligence
The Decision Paradox
The Four Sapiens
Malicious Script of Indian Polity
The IIT Legacy & Global Impact
The IIT Legacy & Global Impact
You are just a Version of Your Original
Win the Game You Didn't Choose
Always Keep Your Bags Packed
Reveal to Shield - Learn the Game of Facades
Engage Beyond Elections
Culture, Identity & Change: The Evolution of Indian Society
The Triangular Dynamics
Never Trust Without Doubt
If You Can't Change the Country, Leave the Country
Live Fearless with Fear
India: The Investment Magnet
Realignment
Powering a Billion Dreams
Universal Skill
Extension: From Brush to Algorithm
Human Side of Einstein
The Future Knows No Lines
Let's Bring the Generations to the Table
Unified Energy Dynamics
A Beat That Never Dies
Strategic Horizons
Right From Scratch

The Forgotten Innovators
Physics Fixing
The Savior Illusion
The Selfish Side
The Age of the Weakest Sapiens
Neural Wars
The Absorption: There Is No Death
Strange Therapist
Bubbles: Become a Drop
Beyond the Blue
Half-Scholar
The Law of Truth
Failure
Philosophy 2.0: A Structural Manifesto
Chaos: A Gentle Guide to the Unresolvable
Self-Help: New Syllabus
Rippling
Game: How to Play the Share Market
Out of Syllabus
Saffronized: The Code Behind the Crown
The Anger
Moksha: The Liberation
Chavan vs Einstein: The Universe Was Never Curved
Rocket Science: The Ripple Gambling
The Cosmos: The Still Universe
The Taste of Silence
Chavan's Post-Materialist Human Model
10-Second Rule
Chavan vs Hawking
It was Never a Wave Signal
Quantum Courtroom: The Trial of a Century

Chavan–Heisenberg Ripple Uncertainty Principle
The Great Disruption
Dark ≠ Hidden
Breath
Calculus of Life
The Earth Doesn't Move
Black Hole Don't Exist
Chavan vs Kant
Naked Truth: Everything and Else
The Moon: Co-Author of Life's Script
World War-3
Aham Brahmasmi
Chavan's Declaration
Light: Neither a Wave, Nor a Photon
Chavan vs. Plato
Good Morning
Myth Wrapped in Math

Watch for more at https://www.sandeepjchavan.com/.

www.ingramcontent.com/pod-product-compliance
Lightning Source LLC
LaVergne TN
LVHW090359210926
848480LV00002B/392

* 9 7 9 8 2 3 1 0 3 9 5 5 5 *